Practice FCC-Type Exams for Radiotelephone Operator's License First Class

Practice FCC-Type Exams for Radiotelephone Operator's License First Class

RICHARD J. SMITH
Applications Engineering Manager

VICTOR F. C. VELEY
Dean of Communications Science
Grantham School of Engineering

Radio Communications and Electronics Instructor
Mount San Antonio College

Consulting Editor
Milton Kaufman
author of
Radio Operator's License Q & A Manual

HAYDEN BOOK COMPANY, INC.
Rochelle Park, New Jersey

Library of Congress Cataloging in Publication Data

Smith, Richard J
Practice FCC-type exams for radiotelephone operator's license, first class.

1. Radiotelephone--Examinations, questions, etc.
2. Radio operators--United States. I. Veley, Victor F. C., joint author. II. Title.
TK6554.5.S53 621.3845'076 77-3321
ISBN 0-8104-5974-4

Printed in the United States of America

3 4 5 6 7 8 9 PRINTING

78 79 80 81 82 83 84 85 YEAR

Preface

Instructors in the communications field have long been aware that practice workbooks, as well as good texts, are necessary if students are going to learn the required subjects successfully enough to pass the FCC examinations. This workbook has been prepared as a learning tool that will serve as a means of self-evaluation for communications technicians preparing for the First Class Radiotelephone Operator's License. It can be used for self-study or for classroom study. It is not intended to replace present text books, but rather to reinforce what has been studied in the text and to prepare the student for his FCC examination.

This workbook has been prepared on the assumption that the student has obtained his Third Class Operator's License, which requires passing Elements 1 and 2, and his Second Class Operator's License, which requires passing Element 3. The workbook contains ten practice tests of 50 questions each for Element 4 (Advanced Radio Telephone).

Answers with study references and discussions of difficult points are found in the back of the book, in addition to an answer sheet merely listing the correct choices for the questions. All of the mathematical problems are individually solved. This workbook was specifically written as a companion to Milton Kaufman's Radio Operator's License Q & A Manual, * but may also be used with any text that follows the FCC Study Guide numbers, or any text whose numbering scheme differs from the FCC Study Guide, although in this case the reference numbers would not be valid.

The student should first study the material for Element 4 from his text. He should then read the introductory material of this workbook before taking Test 1. Convenient spaces are provided for the answers. The student should then use the answer sheet located at the back of the book to correct his test. Any questions missed should be reviewed, using the text and referenced answers. This same procedure is to be followed for Tests 2 through 10.

The FCC passing grade is 75%. The student should average 85% or better in this workbook to achieve the level of confidence needed to take the FCC test. It is suggested that if his average score for Tests 1 through 10 is below 85%, he should repeat his study of his text and try again.

The authors wish to thank Milton Kaufman, their consulting editor, for his review of this book, which greatly contributed to the accuracy of its contents.

Richard J. Smith

Victor F. C. Veley

* Hayden Book Company, Inc., Rochelle Park, N.J.

Taking the FCC Examination

1. Get a full night's sleep. The benefits to be gained by a few extra hours of study are often more than offset by the detrimental effects of fatigue.
2. Don't review or study on the morning of the test. You may lose confidence in your ability and not do so well on the test. Make sure that you leave yourself adequate time to reach the testing center before the doors open.
3. Prepare yourself mentally. Remember that for the Element 4 examination you may miss 12 questions and still pass.
4. Take along only the following items when you go into the FCC examination room (no other items are permitted in the room while an exam is in progress):
    a. Two sharp pencils and one ball-point pen.
    b. Chewing gum or life-savers to subdue tension (no smoking is allowed).
    c. The exact cash amount of any fee that may be required by the FCC.
    d. Your valid Second Class License.
    e. A slide rule or pocket calculator may be used provided they do not have mathematical formulas printed on them.
5. Wear comfortable clothes since you will be sitting for a considerable length of time. (Note that the FCC examiner will not allow you to leave the testing room to go to the restroom.)
6. You will have ample time to take the examination (all day if you wish). No points are given for being the first to complete the examination. Take your time and strive for accuracy.
7. It is suggested that the questions should be answered in the following order:
    a. Rules and Regulations and definitions of terms.
    b. Theory questions.
    c. Problems involving mathematics. If you attempt these first, you may become fatigued and then you may answer some of the easier questions incorrectly.
    d. The first time around, skip all questions that seem difficult, and answer these last. <u>However, make certain that every question has been answered even if you have to guess.</u>
    e. Take time to recheck and make sure that all your answers have been placed in the correct letter space of your choice.
8. Put your thoughts down on the scratch paper provided, and do not attempt to work any mathematical problems in your head. Working these problems out on paper is slower but far more accurate.
9. Some of the questions will be similar to those you have taken in the practice tests. However, there may be subtle differences, so do not attempt to recall the answers given in the practice tests but concentrate on the actual FCC question.

Taking a Multiple Choice Test

1. Directions on the FCC answer sheets will indicate the method of marking your selected answer. As an example, if your selection is (b), make a mark in the space provided as shown:

   (a) | | (b) ■ (c) | | (d) | | (e) | |

   Be sure to follow these directions when taking your FCC examinations. They may use a correction machine to correct your test, and without the proper marking, it may not give you credit for a proper answer.
2. First read the question while covering up the five multiple choice answers, and attempt to find the correct solution. Then read the question again, very carefully, in conjunction with each of the five possible answers. In particular, watch out for expressions such as "not true," "not false," "incorrect," etc.
3. Remember that more than one of the answers may be correct, and that this situation will be covered by still another one of the possible answers, such as "Both (a) and (b) are true" or "All of the above are true." In addition, the first four answers may be false, and the fifth answer stating that "None of the above are true" would therefore be correct.
4. Check every answer carefully to make certain that you have correctly indicated your choice. In other words, if your choice is in fact (b), make sure you have not marked (d). This is a very common type of error with questions that are inaccurately answered in the actual FCC examination.
5. If you answer a question incorrectly in this workbook, use the references to find why the answer as given is correct and why your answer was wrong.

Study Tips

1. Develop good study habits. Provide yourself with a quiet place and plan to study one hour or more every day. Have ready all materials necessary. Use an electronics dictionary or a glossary to look up unfamiliar terms and phrases. You must know the proper definition of words and terms to know what is being asked in a question and to select the proper answer.
2. Develop accuracy in placing your answer on the answer sheet. You would be surprised at the number of students who know the proper answer but place it in the wrong space.
3. Do not try to memorize answers (except possibly for the "Rules and Regulations") but try to understand the background of each question.
4. Use a pencil rather than a pen to solve problems. Keep your mathematical work in a neat order. If you solve a mathematical problem incorrectly, you will then be able to find your errors more easily.
5. Make a thorough review after each test of the questions you have missed, using the text and answer sheets as reference material.

Contents

Element 4, Test 1

1. If the response curve of the IF stages of a communications receiver is found to contain split-phase (double hump) tuning, what action would be taken?
(a) None. Double-humping is desirable in some receivers to provide the wide bandpass necessary for the reception of telephony.
(b) IF bandpass must be increased to provide greater freedom from adjacent channel interference.
(c) IF transformers are overcoupled. The coupling must be reduced to narrow the bandwidth.
(d) The coupling between the local oscillator and the mixer must be reduced.
(e) Both (b) and (d) are true.

(a) | | (b) | | (c) | | (d) | | (e) | |

2. Which of the following conditions would cause loss of the high frequencies during the playback of a properly recorded tape?
(a) Improper azimuth adjustment of the head.
(b) Build-up of oxide or dirt on the playback head.
(c) Incorrect bias current in the tape head.
(d) Tape not passing squarely past the head.
(e) All of the above.

(a) | | (b) | | (c) | | (d) | | (e) | |

3. In the design of the VU meter for broadcast service, which of the listed features is false?
(a) A highly damped meter movement is used.
(b) A logarithmic scale is used for the VU values.
(c) A percent scale is included, with 100 percent coinciding with the zero VU point.
(d) Used with an external 3600-ohm series multiplier resistor, the meter will read zero VU when connected across an impedance of 600 ohms and 1 mW is flowing or there is 0.775 V across the 600-ohm impedance.
(e) Same as in (d) above except that the series multiplier resistor is not used.

(a) | | (b) | | (c) | | (d) | | (e) | |

4. The principal effect of overdriving a crystal unit is:
(a) poor conductivity.
(b) that the solder inside the unit melts.
(c) poor frequency stability.
(d) improved aging.
(e) an increased temperature coefficient.

(a) | | (b) | | (c) | | (d) | | (e) | |

5. 1.6 kW of RMS RF power is delivered to an antenna whose resistance is 43 ohms. What is the peak value of current at the antenna feedpoint?
(a) 4.3 A (b) 6.1 A (c) 8.6 A
(d) 9.2 A (e) 9.7 A

(a) | | (b) | | (c) | | (d) | | (e) | |

6. In the relationship between the power radiated from an antenna and the antenna current, the antenna current is:
(a) directly proportional to the square of the radiated power.
(b) directly proportional to the radiated power.
(c) directly proportional to the square root of the radiated power.
(d) inversely proportional to the square of the radiated power.
(e) There is no simple relationship between the antenna current and the radiated power.

(a) | | (b) | | (c) | | (d) | | (e) | |

7. Which of the following is not true concerning the stereophonic subcarrier used in FM broadcast stations?
(a) The frequency of the stereophonic subcarrier shall be 38 kHz $\pm$4 kHz and shall be amplitude modulated with a frequency band of signals of 50 to 15,000 Hz that contain the left-right signal.
(b) The frequency of the stereophonic subcarrier shall be the second harmonic of the pilot subcarrier and shall be suppressed to less than 1 percent modulation of the main carrier.
(c) The stereophonic subcarrier shall be the second harmonic of the pilot subcarrier and shall cross the time axis with a positive slope simultaneously with each crossing of the time axis by the pilot subcarrier.

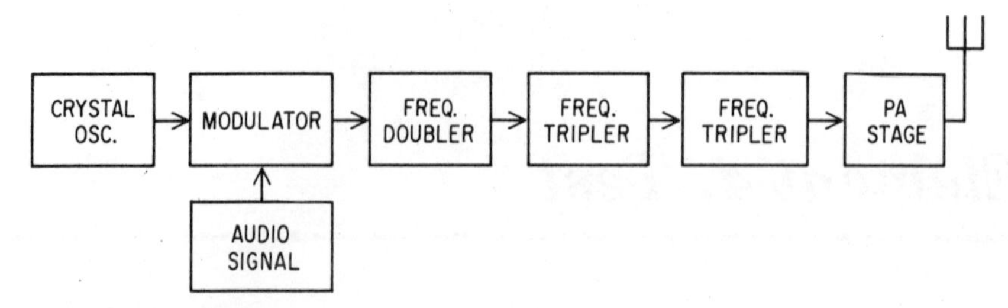

FIGURE 1

(d) The received stereophonic subcarrier is used to directly separate the left and right audio signals at the receiver.
(e) The sum of the sidebands resulting from amplitude modulation of the stereophonic subcarrier shall not cause a peak deviation of the main carrier in excess of 45 percent (excluding the SCA subcarriers).

(a) | | (b) | | (c) | | (d) | | (e) | |

8. In the block diagram shown in Fig. 1, the output frequency is 165.6 MHz. What is the crystal frequency?
(a) 18.4 MHz (b) 9.2 MHz
(c) 27.6 MHz (d) 55.2 MHz
(e) 4.6 MHz

(a) | | (b) | | (c) | | (d) | | (e) | |

9. What is another name for a moving coil microphone?
(a) Crystal microphone
(b) Condenser microphone
(c) Dynamic microphone
(d) Carbon microphone
(e) Ribbon microphone

(a) | | (b) | | (c) | | (d) | | (e) | |

10. A device that can be used to block the RF current in an antenna from entering the power line is:
(a) a diplexer (b) a duplexer
(c) a balun unit (d) a pi-section filter
(e) an Austin ring

(a) | | (b) | | (c) | | (d) | | (e) | |

11. When an H pad is used between a high level AF amplifier and the program transmission line, which of the following will not occur?
(a) A decrease in the program level.
(b) A reduction in the reflections from the line back into the amplifier.
(c) An impedance match.
(d) A reduction in distortion can be obtained.
(e) An increase in the program level.

(a) | | (b) | | (c) | | (d) | | (e) | |

12. What is the output impedance of a crystal microphone?
(a) 50 ohms (b) 100-200 ohms
(c) 600 ohms (d) 1000 ohms
(e) 60-100 kohms

(a) | | (b) | | (c) | | (d) | | (e) | |

13. What minimum class of operator's license is required at a standard broadcast station if its authorized power is 10 kW and the station authorization for its directional antenna requires a current ratio tolerance in the antenna elements of less than 5 percent?
(a) Radiotelephone First Class License.
(b) Radiotelephone Second Class License.
(c) Radiotelephone Third Class License.
(d) Radiotelephone Third Class License with broadcast endorsement.
(e) All the above are true.

(a) | | (b) | | (c) | | (d) | | (e) | |

14. How should antenna structures be painted?
(a) Antennas shall be painted with alternate bands of aviation red and white bands, with a width of approximately one-seventh the height of the structure, and shall be no more than 40 ft nor less than 1 1/2 ft in width.
(b) Antennas shall be painted with alternate bands of aviation orange and white bands, terminating with aviation-surface orange bands at both top and bottom.
(c) Same as in (a) above except that the bands should be no more than 35 ft wide and no less than 2 ft.
(d) Same as in (b) above except that the top and bottom of the tower must terminate in white bands.
(e) Same as in (a) above except that the bands may be no more than 50 ft wide and no less than 5 ft.

(a) | | (b) | | (c) | | (d) | | (e) | |

15. The accuracy of the plate current meter used in the last radio stage of a broadcast transmitter must have a full-scale accuracy of:
(a) 0.01% (b) 0.10% (c) 1.00%
(d) 2.00% (e) 5.00%

(a) | | (b) | | (c) | | (d) | | (e) | |

16. What is the primary reason interlaced scanning is used in television broadcast systems?

(a) Interlaced scanning is the cheapest and easiest system to use.
(b) Interlaced scanning reduces flicker in the picture.
(c) Interlaced scanning improves the vertical linearity of the picture.
(d) Interlaced scanning permits the use of a higher vertical and horizontal oscillator frequency.
(e) Both (a) and (b) are true.

(a) | | (b) | | (c) | | (d) | | (e) | |

17. At its feedpoint, the resistance of an antenna is 120 ohms. If the antenna current is 3.5 A, the RF power at the antenna is:
(a) 1.47 W (b) 147 W (c) 1.47 kW
(d) 420 W (e) 14.7 kW

(a) | | (b) | | (c) | | (d) | | (e) | |

18. Which of the patterns in Fig. 2 is the pattern of the cardioid microphone?

(a) | | (b) | | (c) | | (d) | | (e) | |

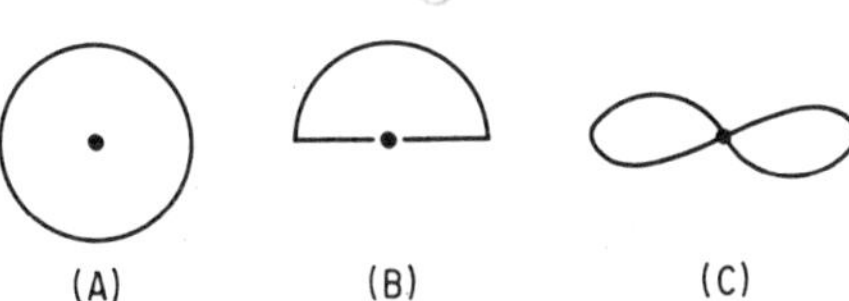

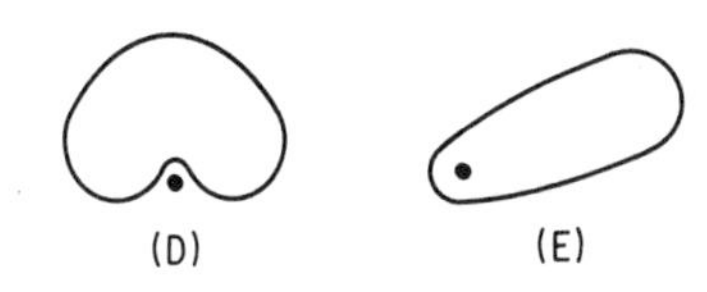

FIGURE 2

19. One method of determining the operating power of the aural transmitter of a TV broadcast station is:
(a) (Antenna current)^2 x Antenna conductance
(b) Plate voltage x Plate current x Efficiency factor F
(c) P_c $(1 + 0.5M^2)$
(d) Antenna power input x (Antenna field gain)^2
(e) Power output from the final stage - Transmission line loss

(a) | | (b) | | (c) | | (d) | | (e) | |

20. Which of the following would not cause a change in the directivity of a directional, two driven element AM broadcast antenna system?
(a) Changes in temperature.
(b) Changes in humidity.
(c) A change in the phase between the two driven elements.
(d) Power line voltage changes to the transmitter.
(e) Changes in the current between the two elements in opposite directions.

(a) | | (b) | | (c) | | (d) | | (e) | |

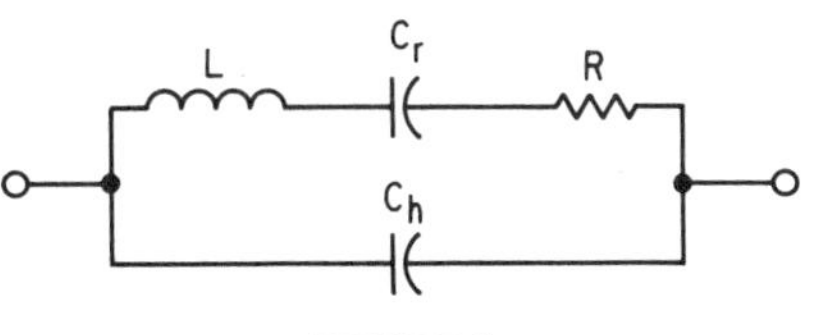

FIGURE 3

21. The circuit shown in Fig. 3 is the equivalent circuit of a:
(a) pentode vacuum tube.
(b) tetrode vacuum tube.
(c) oscillator.
(d) crystal.
(e) None of the above.

(a) | | (b) | | (c) | | (d) | | (e) | |

22. In a series resonant circuit, the ratio of the inductive reactance in ohms to the effective resistance in ohms is called the:
(a) Q of the circuit.
(b) figure of merit of the circuit.
(c) conductance of the circuit.
(d) true power of the circuit.
(e) Both (a) and (b) are true.

(a) | | (b) | | (c) | | (d) | | (e) | |

23. The EBS system was established using commercial broadcasting station facilities in order to:
(a) create a monitoring system for emergency conditions.
(b) disseminate information of an emergency action notification (with or without an attack warning) to licensees and regulated services of the Federal Communications Commission and to the general public during conditions of a grave national crisis or war.
(c) receive distress signals from aircraft or marine services.
(d) provide communication facilities for military messages between service groups.
(e) None of the above is true.

(a) | | (b) | | (c) | | (d) | | (e) | |

24. The purpose of a line equalizer is to:
(a) allow the lower frequencies to pass unattenuated and to attenuate the higher frequency response.
(b) attenuate the lower frequencies and allow the higher frequencies to pass unattenuated to provide a flat frequency response.
(c) reduce the frequency response of the transmission line.
(d) provide a means of impedance match between the telephone line and speech input equipment.
(e) reduce the phase distortion created by the telephone line.

(a) | | (b) | | (c) | | (d) | | (e) | |

25. The term "center frequency":
(a) is the 19-kHz frequency that is located

between the L+R channel and the L-R channel in an FM broadcast signal.
(b) is the frequency of the emitted wave without modulation.
(c) is the average frequency of the emitted wave when modulated by a sinusoidal signal.
(d) is the 38-kHz signal located at the center of the L-R channel of an FM broadcast signal.
(e) Both (b) and (c) are true.

(a) | | (b) | | (c) | | (d) | | (e) | |

26. The output of a program line amplifier is equipped with a bridging type VU meter with a variable attenuator. A 6 VU line pad is connected between the line amplifier and the telephone line. The VU meter indicates -3 VU when the VU attenuator is set at -5 VU. The signal it measures is a 1000-Hz sine wave. What is the power in VU to the telephone line?
(a) +4 VU (b) -2 VU (c) +2 VU
(d) -4 VU (e) Zero VU

(a) | | (b) | | (c) | | (d) | | (e) | |

27. In a triode Class C RF power amplifier under normal operating conditions:
(a) the instantaneous peak positive voltage on the grid must never exceed the corresponding instantaneous voltage on the plate.
(b) the plate current flows when the instantaneous plate voltage is at its highest level.
(c) the plate current does not flow when the instantaneous plate voltage equals the dc plate supply voltage.
(d) the peak surge of plate voltage swing exceeds the value of the plate supply voltage.
(e) Both (a) and (c) are true.

(a) | | (b) | | (c) | | (d) | | (e) | |

28. The dc power input to the final stage of an FM transmitter is 1 kW. This stage has an efficiency of 80 percent, and the transmission line loss is 100 W. If the transmitter is 100 percent modulated by a test tone of 5 kHz and the antenna power gain is 1.5, what is the effective radiated power?
(a) 1675 W (b) 1200 W (c) 1800 W
(d) 1050 W (e) 700 W

(a) | | (b) | | (c) | | (d) | | (e) | |

29. Which of the following expressions represents the power increase produced by a particular stage?
(a) $10 \log_{10} (P_{out}/P_{in})$
(b) $10 \log_{10} (E_{out}/E_{in})$
(c) $10 \log_{10} (I_{out}/I_{in})$
(d) $20 \log_{10} (P_{out}/P_{in})$
(e) Both (b) and (c) are true.

(a) | | (b) | | (c) | | (d) | | (e) | |

30. The frequency tolerance allowed a studio transmitter link (S.T.L.) is:
(a) 0.001% (b) 0.002% (c) 0.003%
(d) 0.004% (e) 0.005%

(a) | | (b) | | (c) | | (d) | | (e) | |

31. Which of the following readings may be recorded in an operating log not to exceed once every three hours?
(a) The final amplifier's plate voltage and plate current.
(b) The reading of the frequency monitor.
(c) The temperature of the crystal oven.
(d) The reading of the antenna current.
(e) All of the above are true.

(a) | | (b) | | (c) | | (d) | | (e) | |

32. To reproduce color accurately, the receiver subcarrier oscillator must be locked in frequency and phase with the transmitters' color subcarrier frequency. This is accomplished by an AFC system utilizing the:
(a) conventional sync and blanking pulses.
(b) "Y" signal. (c) "I" signal.
(d) "Q" signal. (e) Color burst signal.

(a) | | (b) | | (c) | | (d) | | (e) | |

33. When a solid dielectric coaxial transmission line is used to transfer power to an antenna, a sharp kink or bend may cause:
(a) the characteristic impedance of the transmission line to change at that point.
(b) a loss of power through the line.
(c) the SWR to increase.
(d) All of the above are true.
(e) None of the above is true.

(a) | | (b) | | (c) | | (d) | | (e) | |

34. What is the RMS value of a 25 V peak sine wave?
(a) 12.50 V (b) 15.90 V (c) 17.68 V
(d) 21.50 V (e) 22.50 V

(a) | | (b) | | (c) | | (d) | | (e) | |

35. What is the operating power tolerance of an FM broadcast transmitter?
(a) ± 5 percent of the licensed operating power.
(b) ± 10 percent of the licensed operating power.
(c) 90 percent and 105 percent of the licensed operating power.
(d) + 10 percent and -5 percent of the licensed operating power.
(e) + 5 percent and -10 percent of the rated transmitter power.

(a) | | (b) | | (c) | | (d) | | (e) | |

36. When a commercial TV broadcast station is transmitting a video signal, part of the lower and all of the upper sidebands are transmitted. This type of transmission is called:
(a) FSK transmission.
(b) SSB transmission.
(c) SSSC transmission.

(d) Vestigial sideband transmission.
(e) F5 transmission.

(a) | | (b) | | (c) | | (d) | | (e) | |

37. The majority of resonant quartz crystals, when mounted in their holders, form tuned circuits with a Q in the order of:
(a) 1, 000, 000 (b) 200, 000 (c) 20, 000
(d) 1, 000 (e) 100

(a) | | (b) | | (c) | | (d) | | (e) | |

38. Where are pre-amplifiers used in broadcast stations?
(a) Between the microphone and the mixer.
(b) Between the remote line and the mixer.
(c) Between the phone pick-up and the mixer.
(d) Between the tape head and the mixer.
(e) (a), (c), and (d) are true.

(a) | | (b) | | (c) | | (d) | | (e) | |

39. How many horizontal lines are used in the video scanning system used in the United States for TV broadcast stations?
(a) 30 (b) 60 (c) 242.5 (d) 525
(e) 15, 750

(a) | | (b) | | (c) | | (d) | | (e) | |

40. A class C RF final amplifier, using high level modulation, exhibits positive carrier shift during modulation. Which of the following would not cause the trouble?
(a) Parasitic oscillation in the final RF stage.
(b) Insufficient AF excitation to the modulated stage.
(c) Overmodulation.
(d) Excessive negative bias to the class C RF stage.
(e) Improper neutralization of the final RF stage.

(a) | | (b) | | (c) | | (d) | | (e) | |

41. The visual carrier of a broadcast TV station within the United States is:
(a) frequency modulated, and a negative swing of the carrier will be produced by an increase in the picture brilliance.
(b) amplitude modulated, and an increase in picture brilliance will cause an increase in the carrier level.
(c) frequency modulated, and a decrease in picture brilliance will cause the carrier frequency to increase.
(d) amplitude modulated, and a decrease in picture brilliance will cause an increase in the radiated sideband power.
(e) pulse phase modulated, and an increase in the pulse duty time will be caused by a decrease in the picture brilliance.

(a) | | (b) | | (c) | | (d) | | (e) | |

42. The peak-to-peak amplitude of a color burst signal is between:
(a) 90 to 110 percent of the horizontal sync pulses, with the blanking level as a reference.
(b) 50 to 85 percent of the amplitude of the horizontal sync pulse amplitude, with the zero carrier level as a reference.
(c) 70 to 80 percent of the vertical equalizing pulse amplitude, with the black level as a reference.
(d) 80 to 85 percent of the vertical sync pulse amplitude, with the white level as a reference.
(e) 70 to 80 percent of the vertical sync pulse amplitude, with the white level as a reference.

(a) | | (b) | | (c) | | (d) | | (e) | |

43. Under static conditions, a transistor's β is 50, and the collector current is 6 mA. The emitter current is:
(a) 6.12 mA (b) 300 mA (c) 0.12 mA
(d) 5.88 mA (e) 5.76 mA

(a) | | (b) | | (c) | | (d) | | (e) | |

44. An AM transmitter has a carrier power output of 20 kW and is modulated 90 percent by a sinusoidal 2.5 kHz audio frequency. What is the power in each sideband?
(a) 2.00 kW (b) 2.50 kW (c) 4.05 kW
(d) 7.50 kW (e) 8.1 kW

(a) | | (b) | | (c) | | (d) | | (e) | |

45. An AM broadcast station receives a report that, at the time its carrier frequency was checked, it was 16 Hz high. The transmitter's log for the same time indicates that the carrier frequency was 10 Hz low. What is the error in the station's frequency monitor?
(a) +26 Hz (b) -26 Hz (c) +6 Hz
(d) -6 Hz (e) -10 Hz

(a) | | (b) | | (c) | | (d) | | (e) | |

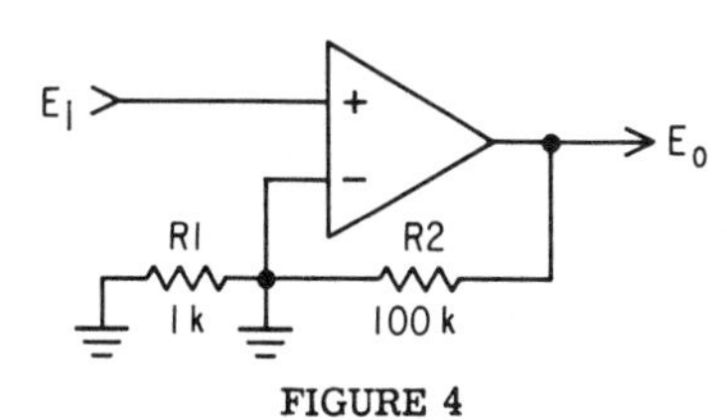

FIGURE 4

46. Referring to the operational amplifier circuit of Fig. 4, which of the following statements is false?
(a) The input impedance is 1 kohm.
(b) The signal gain $E_0/E1$ is 101.
(c) The voltage at the inverting and non-inverting inputs of the operational amplifier are virtually the same.
(d) The high frequency cutoff frequency is determined by the high frequency characteristics of the operational amplifier and the resistors R1 and R2.
(e) The phase shift of the circuit at low frequencies is 0°.

(a) | | (b) | | (c) | | (d) | | (e) | |

47. The input frequency to a push-push doubler stage is 1.7 MHz. The capacitance of the plate tank circuit is 64 pF. What is the value of the plate tank coil?
(a) 16.9 μH (b) 34.3 μH (c) 169 μH
(d) 67.6 μH (e) 338 μH

(a) | | (b) | | (c) | | (d) | | (e) | |

48. What is the height of a vertical radiator, one-quarter wavelength long, that is operated at a frequency of 1.5 MHz?
(a) 47.0 m (b) 50.0 m (c) 77.0 m
(d) 82.0 m (e) 188.0 m

(a) | | (b) | | (c) | | (d) | | (e) | |

49. An AM transmitter is high level modulated by a single audio tone. If the modulation percentage is reduced from 80 to 60 percent, what is the percentage reduction in total sideband power?
(a) 32% (b) 18% (c) 44% (d) 78%
(e) 15%

(a) | | (b) | | (c) | | (d) | | (e) | |

50. In a purely capacitive circuit, the voltage:
(a) leads the current by 90°.
(b) leads the current by 45°.
(c) lags the current by 90°.
(d) lags the current by 45°.
(e) and the current are in phase.

(a) | | (b) | | (c) | | (d) | | (e) | |

Element 4, Test 2

1. Regarding a quartz crystal, which of the following statements is false?
(a) The type of cut determines the mode of operation.
(b) A crystal may be cut to produce a large number of strong overtones.
(c) A crystal behaves inductively between its series and parallel resonant frequencies.
(d) The fundamental frequency is directly proportional to the thickness of the crystal slice.
(e) A crystal may have an equivalent Q of several thousand.

(a) | | (b) | | (c) | | (d) | | (e) | |

2. What is the bandwidth of a standard broadcast TV channel?
(a) 4.2 MHz (b) 4.5 MHz (c) 6.0 MHz
(d) 10.0 MHz (e) 12.0 MHz

(a) | | (b) | | (c) | | (d) | | (e) | |

3. In Fig. 1, the emitter-base junction voltage drop is 0.3 V and the static current gain (α) is 0.95. The value of the collector current is:
(a) 0.42 mA (b) 0.44 mA (c) 0.46 mA
(d) 0.48 mA (e) 0.5 mA

(a) | | (b) | | (c) | | (d) | | (e) | |

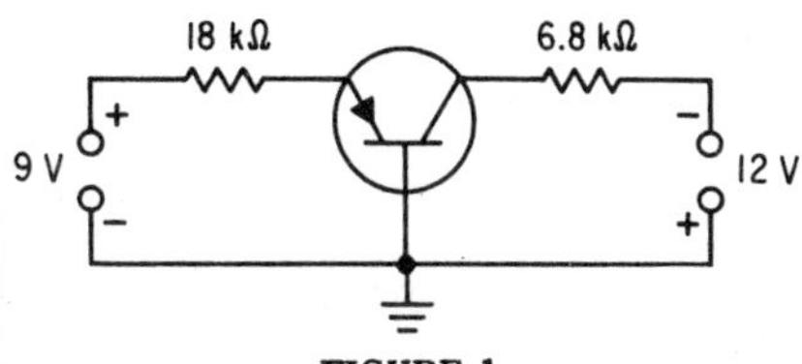

FIGURE 1

4. In Fig. 1, the collector potential is:
(a) +6.8 V (b) -7.9 V (c) -8.9 V
(d) -9.6 V (e) -10.4 V

(a) | | (b) | | (c) | | (d) | | (e) | |

5. A series resonant crystal oscillator is defined as an oscillator:
(a) in which all the components are in series.
(b) in which one side of the crystal is grounded.
(c) which has minimum impedance and zero phase shift at the crystal terminals.
(d) which has high impedance.
(e) with a capacitance placed in series with the crystal.

(a) | | (b) | | (c) | | (d) | | (e) | |

6. What portion of a square law current meter scale is used for accuracy?
(a) Lower half. (b) Upper half.
(c) Lower two-thirds. (d) Upper two-thirds.
(e) Upper one-third.

(a) | | (b) | | (c) | | (d) | | (e) | |

7. If it is desired to conduct program tests on an AM Broadcast Station, what is the maximum number of days prior to the conducting of the tests that the request must be filed with the FCC?
(a) 2 days. (b) 5 days. (c) 10 days.
(d) 20 days. (e) 30 days.

(a) | | (b) | | (c) | | (d) | | (e) | |

8. An antenna tower is electrically 0.32 λ long at the operating frequency of 1460 kHz. What is the antenna's physical length in meters?
(a) 62.5 (b) 182.4 (c) 165.7 (d) 97.4
(e) 194.9

(a) | | (b) | | (c) | | (d) | | (e) | |

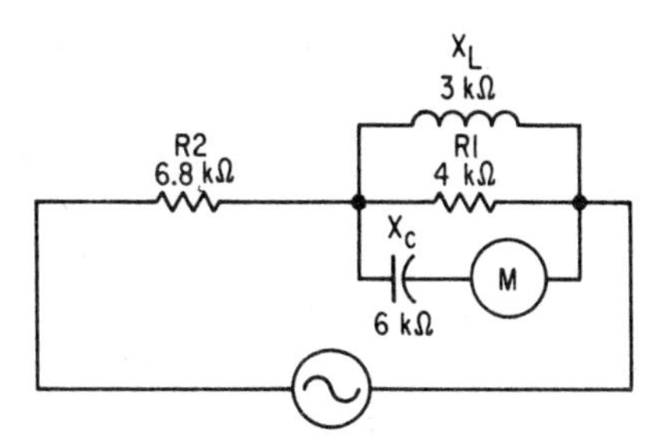

FIGURE 2

9. In the circuit shown in Fig. 2, meter M reads 12 mA. Calculate the voltage drop across R_2.

147 volts.

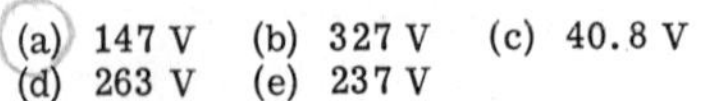

(a) 147 V (b) 327 V (c) 40.8 V
(d) 263 V (e) 237 V

(a) | | (b) | | (c) | | (d) | | (e) | |

10. A balanced microphone cable has:
(a) one conductor and one shield.
(b) two conductors and one shield.
(c) three conductors and one shield.
(d) three conductors.
(e) four conductors.

(a) | | (b) | | (c) | | (d) | | (e) | |

11. How often must Off-The-Air Monitor EBS Tests be transmitted?
(a) Once per day. (b) Twice per week.
(c) Once per week. (d) Twice per month.
(e) Once per month.

(a) | | (b) | | (c) | | (d) | | (e) | |

12. In TV broadcast stations, what is the horizontal scanning frequency?
(a) 30 Hz (b) 60 Hz (c) 525 Hz
(d) 15, 750 Hz (e) 31, 500 Hz

(a) | | (b) | | (c) | | (d) | | (e) | |

13. In the differential amplifier circuit shown in Fig. 3, E1 and E2 are shorted together, and a dc meter reading from E_0 to ground indicates a +0.2 V output. A possible cause is:
(a) potentiometer R5 is misadjusted.
(b) R2 is an open circuit.
(c) R1 is an open circuit.
(d) R3 is an open circuit.
(e) R4 is an open circuit.

(a) | | (b) | | (c) | | (d) | | (e) | |

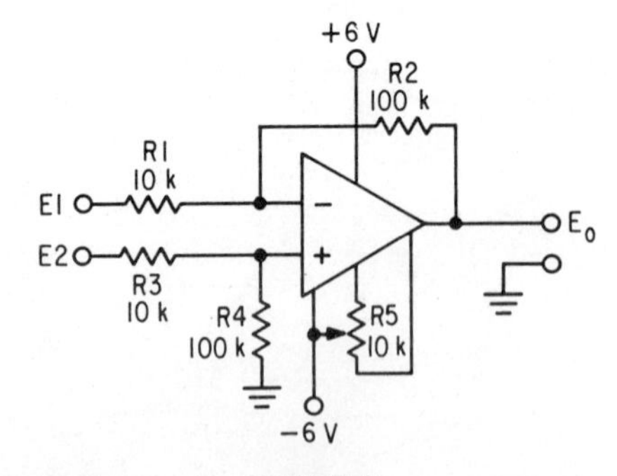

FIGURE 3

14. What type of circuit is represented in Fig. 4?
(a) A triangular wave generator.
(b) A sawtooth wave generator.
(c) A square wave generator.
(d) A pulse wave generator.
(e) A sinusoidal wave generator.

(a) | | (b) | | (c) | | (d) | | (e) | |

15. The circuit shown in Fig. 4 could be used in broadcast television stations:

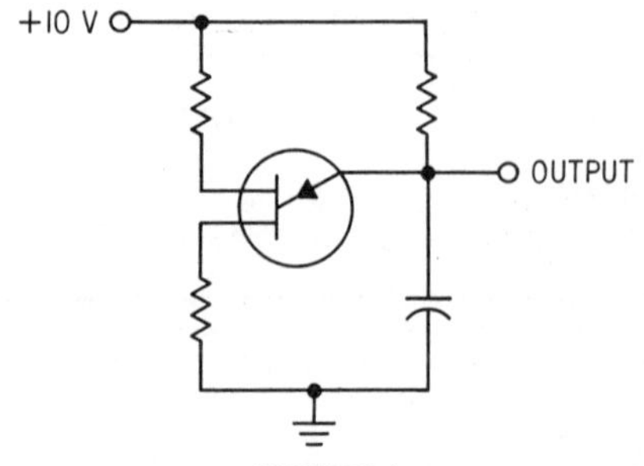

FIGURE 4

(a) to control the carrier frequency of the video transmitter.
(b) in the TV cameras to control the trace and re-trace of the electron beam.
(c) as an automatic device to control the brilliance of the picture.
(d) in ocilloscopes to generate the horizontal sweep.
(e) Both (b) and (d) are true.

(a) | | (b) | | (c) | | (d) | | (e) | |

16. Which two record speeds are not in common use?
(a) 16 and 33 1/3 rpm (b) 45 and 78 rpm
(c) 33 1/3 and 45 rpm (d) 33 1/3 and 78 rpm
(e) (a), (b), and (d) are all correct.

(a) | | (b) | | (c) | | (d) | | (e) | |

17. The main channel input of an FM broadcast stereophonic transmission shall contain signals that:
(a) include the frequencies of 50 to 15, 000 Hz
(b) include the sum of the left and right signals.
(c) include the difference of the left and right signals.
(d) are used to control and separate the stereo signals at the receiver.
(e) Both (a) and (b) are true.

(a) | | (b) | | (c) | | (d) | | (e) | |

18. "Shot effect" in diodes is caused by:
(a) random electron movement from cathode to the plate.
(b) excessive plate current.
(c) interelectrode capacitance.
(d) noise.
(e) radiation.

(a) | | (b) | | (c) | | (d) | | (e) | |

19. An AM transmitter is high level modulated by a single audio tone. If the percentage of modulation is changed from 60 to 80 percent, what is the percentage increase in antenna current?
(a) 4.5% (b) 5.5% (c) 20% (d) 5.2%
(e) 10%

(a) | | (b) | | (c) | | (d) | | (e) | |

20. Frequency modulation of the tones on a record caused by the mechanics of the turntable is known as:
(a) frequency response. (b) drag
(c) flutter. (d) RMS.
(e) skating.

(a) | | (b) | | (c) | | (d) | | (e) | |

21. An audio amplifier uses a PNP transistor. If the voltage on the collector with respect to the emitter is negative while the voltage on the base with respect to the emitter is positive:
(a) conditions are normal.
(b) the voltage on the collector is correct but the voltage on the base is wrong.
(c) the voltage on the collector is wrong but the voltage on the base is correct.
(d) both voltages are wrong.
(e) the transistor is being used as part of a complementary push-pull circuit.

(a) | | (b) | | (c) | | (d) | | (e) | |

22. A series resonant LCR circuit includes a resistor of 400 ohms and an inductive reactance of 160 ohms. Assuming that the frequency is unchanged, what will be the total impedance of the circuit if the inductance is tripled and the capacitance is doubled?
(a) 800 ohms (b) 960 ohms (c) 566 ohms
(d) 640 ohms (e) 780 ohms

(a) | | (b) | | (c) | | (d) | | (e) | |

23. For standard AM broadcast stations, the maximum modulation percentage on negative peaks is:
(a) 50% (b) 65% (c) 75% (d) 85%
(e) 100%

(a) | | (b) | | (c) | | (d) | | (e) | |

24. Figure 5 represents a:
(a) low-pass filter. (b) high-pass filter.
(c) line equalizer. (d) delay line.
(e) Both (b) and (c) are true.

(a) | | (b) | | (c) | | (d) | | (e) | |

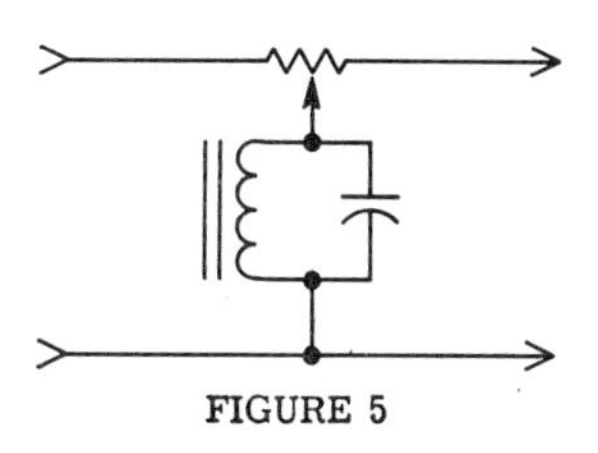

FIGURE 5

25. Which of the following is required to be entered in the TV station's maintenance log?
(a) The operating contents of the last radio stage of the aural transmitter.
(b) An entry each week of the time and result of tests of the auxiliary transmitter, if installed.
(c) Transmission line meter readings for both visual and aural transmitters.
(d) For remote control operation, the results of observations of the vertical interval test signal transmissions.
(e) An entry of the time the station begins to supply power to the antenna and the time it stops.

(a) | | (b) | | (c) | | (d) | | (e) | |

26. Due to a mismatch between a transmission line and its load, 10 percent of the power arriving at the load is reflected back to the source. What is the value of the voltage standing wave ratio on the line?
(a) 1.316 (b) 1.92 (c) 0.76 (d) 1.25
(e) 2.374

(a) | | (b) | | (c) | | (d) | | (e) | |

27. The video transmitter carrier frequency of a broadcast television station must be maintained within what tolerance of its authorized carrier frequency?
(a) ± 20 Hz (b) ± 100 kHz (c) ± 1000 Hz
(d) ± 2000 Hz (e) ± 75 kHz

(a) | | (b) | | (c) | | (d) | | (e) | |

28. The field intensity at a position three miles away from a transmitter is 80 mV per meter. How far from the transmitter will be the 50 mV per meter contour if the transmitter power is doubled?
(a) 9.6 miles (b) 4.8 miles
(c) 6.8 miles (d) 7.2 miles
(e) 8.4 miles

(a) | | (b) | | (c) | | (d) | | (e) | |

29. How often must the proof-of-performance measurements be made on Standard Broadcast and FM broadcast stations?
(a) Daily. (b) Weekly.
(c) Monthly. (d) Semiannually.
(e) At reasonable intervals throughout the year.

(a) | | (b) | | (c) | | (d) | | (e) | |

30. The two-tube circuit shown in Fig. 6 represents:

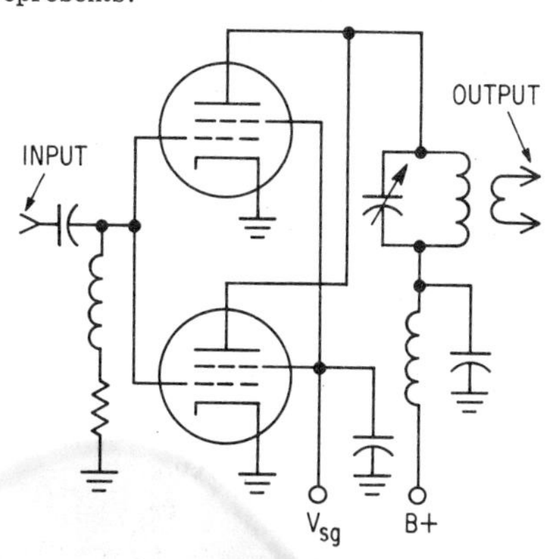

FIGURE 6

R-F PWR AMPL. with two tubes connected in parallel.

(a) an RF power amplifier with two tubes connected in parallel.
(b) an RF power amplifier with two tubes connected in push-pull.
(c) an RF power amplifier with two tubes connected in push-push.
(d) a reactance tube modulator.
(e) a plate neutralized RF amplifier.

(a) | | (b) | | (c) | | (d) | | (e) | |

31. In a series resonant circuit the resistance, the inductive reactance, and the capacitive reactance are each 120 ohms. If the inductance is doubled but the capacitance is halved, what will be the new impedance of the circuit?
(a) 60 ohms (b) 120 ohms (c) 240 ohms
(d) 30 ohms (e) 480 ohms

(a) | | (b) | | (c) | | (d) | | (e) | |

32. Time delay relays may be used in broadcast transmitters to:
(a) keep the carrier on the air during periods of transitory overload.
(b) provide a period of time for the mercury-vapor rectifier filaments to warm up before the plate voltage is applied.
(c) provide a period of time for the filaments of high powered RF and AF vacuum tubes to warm up before the plate voltage is applied.
(d) keep the transmitter on the air during short periods of overmodulation.
(e) Both (b) and (c) are true.

(a) | | (b) | | (c) | | (d) | | (e) | |

33. An audio amplifier has a circuit that provides degenerative feedback in the amplifier. This will cause:
(a) less amplifier gain, but will improve the fidelity.
(b) an increase in gain and distortion.
(c) a reduction in gain, and an increase in distortion.
(d) self oscillation.
(e) an increase in gain, and a reduction in distortion.

(a) | | (b) | | (c) | | (d) | | (e) | |

34. If a relay contact becomes pitted, which of the following methods should be used to repair it?
(a) Remove the roughness with a thin strip of sandpaper to produce a flat smooth surface for the point contact area; then clean with a lint-free cloth.
(b) Remove the roughness with a metal file to produce a crowned shape on the contact points; then clean with a lint-free cloth.
(c) Use a burnishing tool to repair badly pitted contacts; then clean with a contact cleaner.
(d) Remove the roughness with a clean metal file, taking off as little of the contact material as possible while maintaining the original contact shape.
(e) None of the above is true for if the relay contacts become pitted, the relay must be replaced.

(a) | | (b) | | (c) | | (d) | | (e) | |

35. A three-quarter wavelength section of transmission line is terminated by a resistive load of 125 ohms. The surge impedance of the line is 70 ohms. What is the input impedance to the line?
(a) 223 ohms (b) 195 ohms (c) 93.4 ohms
(d) 46.7 ohms (e) 39.2 ohms

(a) | | (b) | | (c) | | (d) | | (e) | |

36. In what direction should a light-sensitive device used to control tower lights face?
(a) North. (b) South. (c) East. (d) West.
(e) It should face in the direction as noted in the station license.

(a) | | (b) | | (c) | | (d) | | (e) | |

37. A pentode audio amplifier uses a tube with a transconductance of 3,500 micromhos. The plate load resistance is 22 kohms, the grid bias is -1.5 V, the plate current is 4 mA and the plate supply voltage is 250 V. What is the voltage gain of the amplifier?
(a) 88 (b) 166 (c) 77 (d) 59 (e) 44

(a) | | (b) | | (c) | | (d) | | (e) | |

38. The mutual inductance between two coils is 5 mH. The inductances of the two coils are 0.3 H and 15 mH. What is the coefficient of coupling between the two coils?
(a) 0.055 (b) 0.0745 (c) 0.745
(d) 1.111 (e) 1.59

(a) | | (b) | | (c) | | (d) | | (e) | |

39. The fourth harmonic of 1.45 MHz is:
(a) 1.45 MHz (b) 2.90 MHz (c) 4.35 MHz
(d) 5.80 MHz (e) 7.25 MHz

(a) | | (b) | | (c) | | (d) | | (e) | |

40. The decelerator grid of the image-orthicon television camera tube is used to:
(a) cause the electron beam to scan the target area.
(b) slow the electron beam before it reaches the target.
(c) focus the electron beam on the photo-cathode.
(d) keep the electrons in the scanning beam from bunching.
(e) improve the focus of the scanning beam.

(a) | | (b) | | (c) | | (d) | | (e) | |

41. Overmodulation is indicated in an AM broadcast station by which of the following oscilloscope patterns shown in Fig. 7?

(a) | | (b) | | (c) | | (d) | | (e) | |

42. Referring to the operational amplifier circuit of Fig. 8, which of the following statements is false?
(a) The input impedance is 1 kohm.
(b) The signal gain E_O/E_I is +100.

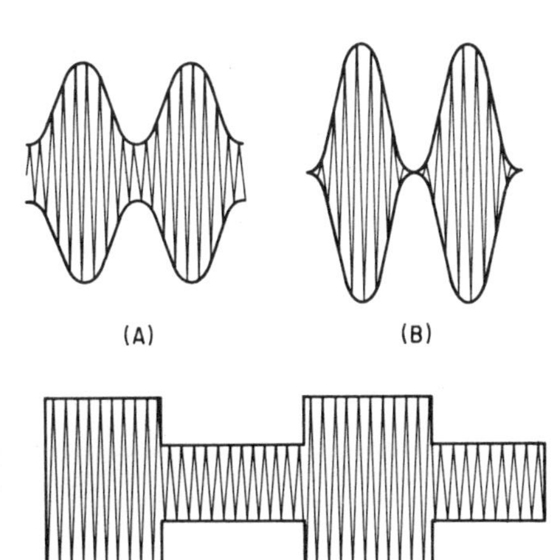

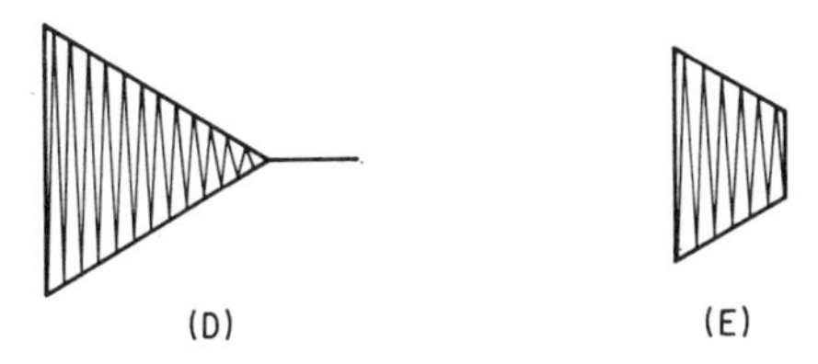

FIGURE 7

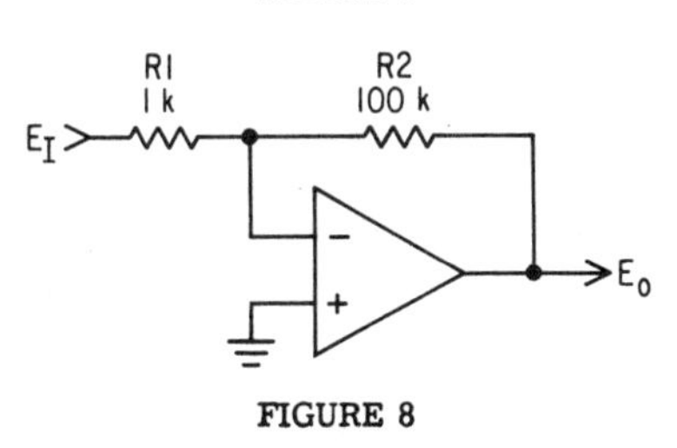

FIGURE 8

(c) The signal voltage at the inverting input of the operational amplifier is virtually 0 V.
(d) The high frequency cutoff frequency is determined by the high frequency characteristics of the operational amplifier and the resistors R1 and R2.
(e) The phase shift of the circuit at low frequencies is 180°.

(a) | | (b) | | (c) | | (d) | | (e) | |

43. Which of the following is true concerning the pilot subcarrier frequency, tolerance, and percent modulation used with an FM stereophonic broadcast station?
(a) The pilot subcarrier frequency is 38 kHz ± 2 Hz, and it amplitude modulates the main carrier frequency between 8 and 10 percent.
(b) The pilot subcarrier frequency is 19 kHz ± 10 Hz, and it frequency modulates the main carrier between 10 and 20 percent.
(c) The pilot subcarrier frequency is 19 kHz ± 2 Hz, and it frequency modulates the main carrier between 8 and 10 percent.
(d) The pilot subcarrier frequency is 38 kHz ± 10 Hz, and it frequency modulates the main carrier between 10 and 20 percent.
(e) The pilot subcarrier frequency is 19 kHz ± 2 Hz, and it frequency modulates the main carrier between 45 and 50 percent.

(a) | | (b) | | (c) | | (d) | | (e) | |

44. The frequency tolerance allowed an international broadcast station is:
(a) 0.001% (b) 0.0015% (c) 0.003%
(d) 0.004% (e) 0.005%

(a) | | (b) | | (c) | | (d) | | (e) | |

45. Why is "top loading" sometimes used for standard broadcast station antenna systems?
(a) To increase the effective length of the antenna.
(b) To lower the angle of radiation.
(c) To improve the current distribution of the antenna.
(d) To increase the angle of radiation.
(e) (a), (b), and (c) are true.

(a) | | (b) | | (c) | | (d) | | (e) | |

46. The maximum allowable carrier shift of a standard broadcast transmitter is:
(a) 5% (b) 2.5% (c) 10% (d) -10%, + 5%
(e) +10%, - 5%

(a) | | (b) | | (c) | | (d) | | (e) | |

47. Which of the following statements is not true?
(a) Some models of the Vidicon television camera tubes have greater resolution capability than the Image-Orthicon tube.
(b) Some models of the Vidicon television camera tubes are preferred for portable camera operation.
(c) Most models of the Vidicon television require less input power.
(d) Some models of the Image-Orthicon tube are more rugged than Vidicon camera tubes and are preferred for portable operation.
(e) Both (a) and (b) are false statements.

(a) | | (b) | | (c) | | (d) | | (e) | |

48. A factor used for determining the operating power of a standard broadcast station by the indirect method is designated by the letter:
(a) P (b) R (c) F (d) G (e) Z

(a) | | (b) | | (c) | | (d) | | (e) | |

49. The chrominance subcarrier frequency is located on the:

(a) "front porch" of the horizontal sync pulses and consists of eight or more cycles at a frequency of 15.75 kHz.
(b) "front porch" of the horizontal sync pulses and consists of five or more cycles at a frequency of 3.579545 MHz.
(c) "back porch" of the horizontal blanking pulses and consists of eight or more cycles at a frequency of 3.579545 MHz.
(d) "back porch" of the vertical sync pulses and consists of eight or more cycles at a frequency of 15.75 MHz.
(e) "back porch" of the vertical equalizing pulse interval at a frequency of 3.579545 MHz.

(a) | | (b) | | (c) | | (d) | | (e) | |

50. If a certain power supply has an output voltage of 230 V at full load and a regulation of 22 percent, what is the output voltage with no load?
(a) 230 V (b) 250 V (c) 270 V
(d) 281 V (e) 269.5 V

(a) | | (b) | | (c) | | (d) | | (e) | |

Element 4, Test 3

1. A VU meter with a 6 VU T pad in the meter circuit is monitoring a program line transmission of music. The meter indicates that the program level is +1 VU. The actual program level is:
(a) +5 VU RMS (b) +7 VU peak
(c) +5 VU peak (d) +7 VU average
(e) +7 VU peak-to-peak

(a) | | (b) | | (c) | | (d) | | (e) | |

2. Refer to Fig. 1. E1 and E2 are tied together, and an ac signal is placed between them and ground. E_0 indicates zero output. What, if anything, is the cause?
(a) R4 is an open circuit.
(b) R1 is an open circuit.
(c) Nothing is wrong.
(d) R3 is an open circuit.
(e) R2 is an open circuit.

(a) | | (b) | | (c) | | (d) | | (e) | |

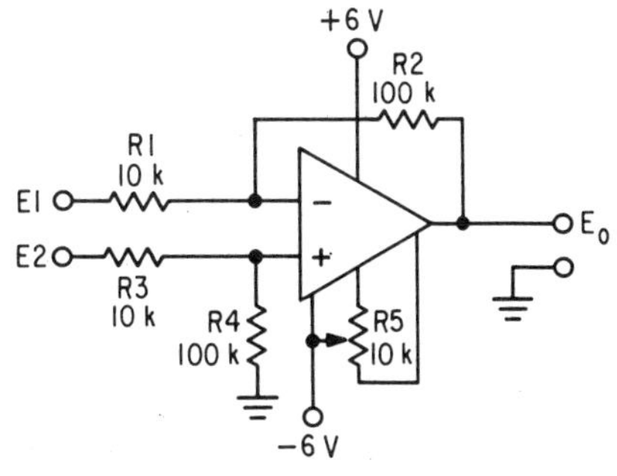

FIGURE 1

3. The circuit shown in Fig. 2 employs:
(a) regenerative feedback.
(b) current degenerative feedback.
(c) voltage degenerative feedback.
(d) a treble tone control.
(e) a bass tone control.

(a) | | (b) | | (c) | | (d) | | (e) | |

4. When test signals are transmitted along with program material to check the quality of the transmission system of a broadcast TV station, they are inserted:
(a) between the seventeenth through the twentieth line of the vertical blanking interval of each field.
(b) on the front porch.
(c) on the back porch.
(d) during the horizontal pulses.
(e) at the beginning and end of each horizontal line.

(a) | | (b) | | (c) | | (d) | | (e) | |

5. In a television receiver, synchronizing pulses are fed to the:
(a) high voltage output stage.
(b) automatic volume control circuit.
(c) vertical oscillator and horizontal oscillator AFC stages.
(d) delay line.
(e) damper stage.

(a) | | (b) | | (c) | | (d) | | (e) | |

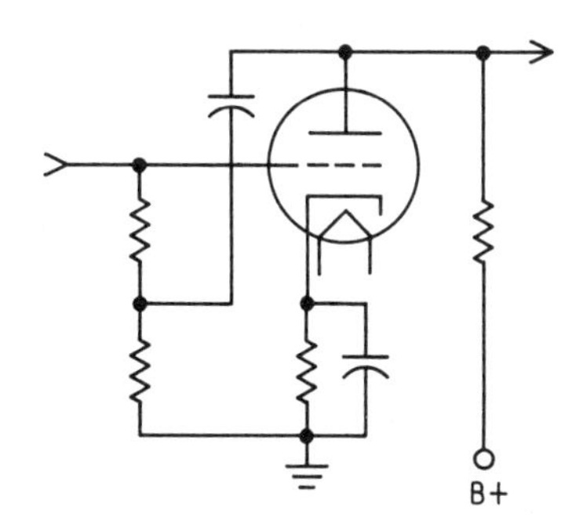

FIGURE 2

6. What is the purpose of R1 and C2 in the class C RF power amplifier in Fig. 3?
(a) They provide "grid neutralization" for the stage.
(b) They improve the Q of the tuned grid circuit.
(c) They provide class C bias for the stage.
(d) They prevent parasitic oscillations.
(e) They increase the voltage gain of the stage.

(a) | | (b) | | (c) | | (d) | | (e) | |

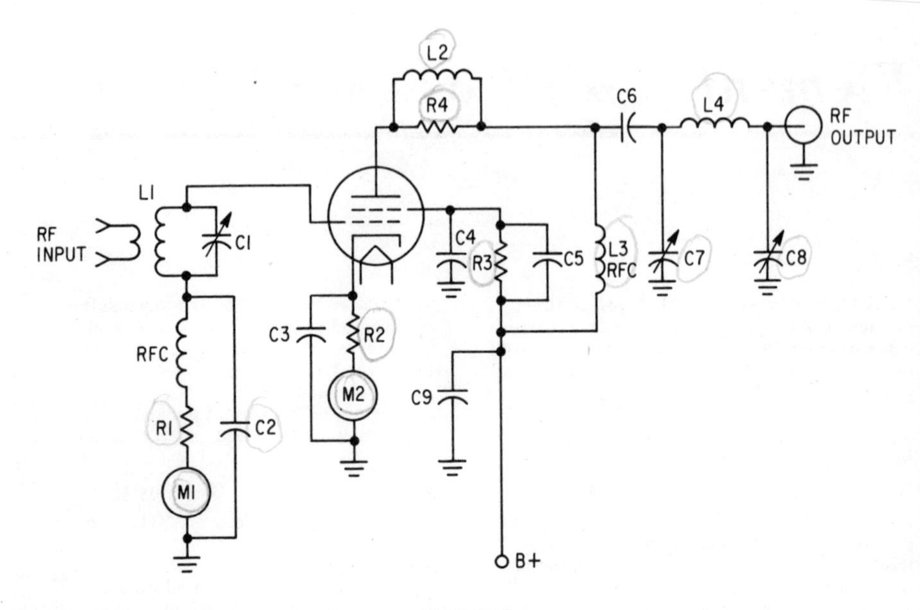

FIGURE 3

7. In the schematic of Fig. 3, what is the purpose of L2 and R4?
(a) L2 and R4 are used for inductive neutralization for the stage.
(b) They provide parasitic suppression for the stage.
(c) L2 and R4 form a peaking circuit to improve the high frequency response of the stage.
(d) L2 and R4 provide an output load for the stage.
(e) None of the above is true.

(a) | | (b) | | (c) | | (d) | | (e) | |

8. In the schematic of Fig. 3, what is the purpose of L3?
(a) L3 permits the plate to be shunt fed.
(b) L3 acts as an antenna matching coil.
(c) L3 is used for parasitic suppression.
(d) L3 provides inductive neutralization for the stage.
(e) L3 is part of the resonant plate tank circuit.

(a) | | (b) | | (c) | | (d) | | (e) | |

9. In the schematic of Fig. 3, what is the purpose of C7, C8, and L4?
(a) They form a pre-emphasis circuit.
(b) They provide an impedance match between the plate of the tube and the output.
(c) They form a T network.
(d) They form a Pi network.
(e) Both (b) and (d) are true.

(a) | | (b) | | (c) | | (d) | | (e) | |

10. In the schematic of Fig. 3, if the RF excitation is lost, what component part or parts would protect the tube from excessive plate current?
(a) R2 (b) R1 (c) R3 (d) L2 and R4
(e) L3

(a) | | (b) | | (c) | | (d) | | (e) | |

11. In the schematic of Fig. 3, if the RF excitation is lost, what would happen to the meter readings of M1 and M2?
(a) The readings would be unchanged.
(b) M1 would fall to zero, and M2 would increase.
(c) M1 would increase and M2 would decrease
(d) Both readings would increase.
(e) Both readings would decrease.

(a) | | (b) | | (c) | | (d) | | (e) | |

12. In the schematic of Fig. 3, what change, if any, would be indicated on meter M2 if R3 were open?
(a) The reading would become very low or zero.
(b) The reading would increase.

(c) The reading would be off scale high.
(d) The reading would be off scale low.
(e) No change would be noticed.

(a) | | (b) | | (c) | | (d) | | (e) | |

13. Two vertical broadcast station antennas are 180° apart, and antenna B is driven 180° out of phase with antenna A. With equal power being radiated by each antenna, in which direction is the maximum radiation?
(a) The pattern would be hemispherical in the vertical plane and circular in the horizontal plane.
(b) At right angles to a line bisecting the two antennas.
(c) The radiation pattern would be perfectly circular.
(d) Along the line bisecting the antennas.
(e) At 45° to a line bisecting the two antennas.

(a) | | (b) | | (c) | | (d) | | (e) | |

14. Linear RF amplifiers are usually operated in class B instead of class A because:
(a) the efficiency of the former is approximately 35 percent.
(b) remote cut-off type tubes can be used.
(c) as a consequence of the higher efficiency of the former, they require less plate supply power for a given output power.
(d) class B produces less distortion of the signal than class A.
(e) Both (a) and (c) are true.

(a) | | (b) | | (c) | | (d) | | (e) | |

15. Carrier shift may be caused in an AM broadcast transmitter by:
(a) overmodulation.
(b) an incorrect load impedance presented to the modulator tube by the RF amplifier.
(c) poor regulation of the RF amplifier power supply.
(d) improper tuning of the plate tank circuit.
(e) All of the above are true.

(a) | | (b) | | (c) | | (d) | | (e) | |

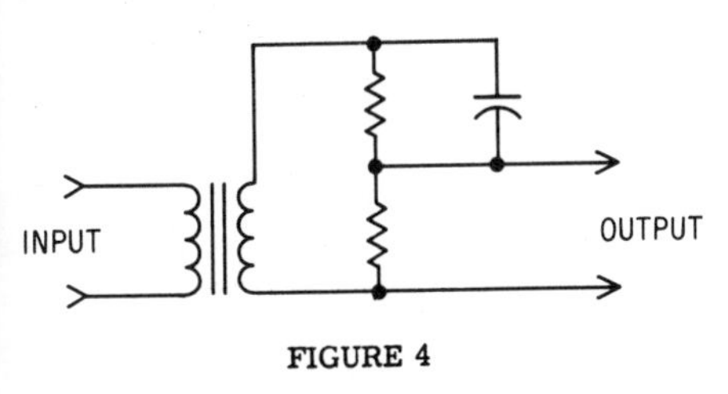

FIGURE 4

16. The circuit shown in Fig. 4 represents:
(a) an audio tone control network.
(b) a de-emphasis circuit.
(c) a pre-emphasis circuit.
(d) an audio correction network for converting AM to FM.
(e) an impedance matching network for connecting the microphone to the preamplifier.

(a) | | (b) | | (c) | | (d) | | (e) | |

17. The time constant of the pre-emphasis CR network used for FM broadcast is:
(a) 50 μ seconds. (b) 75 μ seconds.
(c) 100 μ seconds. (d) 150 μ seconds.
(e) not constant, but varies with the frequency of the audio signal.

(a) | | (b) | | (c) | | (d) | | (e) | |

18. The term "crystal correlation" refers to:
(a) the crystal holder dimensions.
(b) oscillator component values in terms of their use as a function of the Q of the crystal.
(c) the crystal dimensions.
(d) the type of crystal to be used.
(e) the establishing of test and operating conditions, such as load capacitance, operating temperature, and drive level.

(a) | | (b) | | (c) | | (d) | | (e) | |

19. The inductance of a crystal in its equivalent circuit is:
(a) related to the mass of the crystal.
(b) one half of the Q.
(c) the mechanical elasticity of the quartz.
(d) the molecular friction of the quartz.
(e) the primary factor in determining the crystal drive level.

(a) | | (b) | | (c) | | (d) | | (e) | |

20. The frequency range of an SCA subcarrier when stereophonic transmission is not in use must be within the range of:
(a) 20 to 75 kHz (b) 53 to 75 kHz
(c) 20 to 53 kHz (d) 50 Hz to 15 kHz
(e) 20 Hz to 20 kHz

(a) | | (b) | | (c) | | (d) | | (e) | |

21. What percentage of the peak carrier level, as represented by the synchronizing pulses, is the blanking level, as required by the FCC?
(a) 12.5% (b) 25% (c) 50% (d) 75%
(e) 80%

(a) | | (b) | | (c) | | (d) | | (e) | |

22. Which of the following types of emission do AM standard broadcast stations use?
(a) A1 (b) A2 (c) A3 (d) A5A (e) F3

(a) | | (b) | | (c) | | (d) | | (e) | |

23. Referring to the operational amplifier circuit of Fig. 5, which of the following is false?
(a) The low frequency gain, $E_o/E1$, is +11.
(b) The input impedance is very high.
(c) The capacitor C1 provides control of the high frequency cutoff frequency.
(d) The high frequency gain is +10.

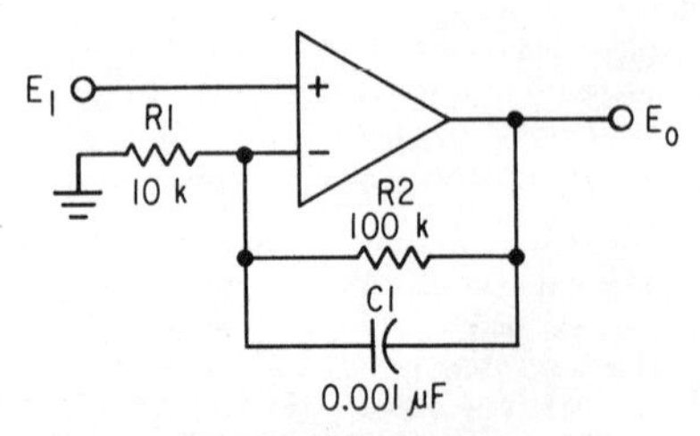

FIGURE 5

(e) The high frequency gain approaches zero ($E_0/E1$).

(a) | | (b) | | (c) | | (d) | | (e) | |

24. Which of the following devices are used in broadcast transmitters for the safety of personnel?
(a) Parasitic suppressors.
(b) Overload relays.
(c) Bleeder resistors.
(d) Interlocks.
(e) Both (c) and (d) are true.

(a) | | (b) | | (c) | | (d) | | (e) | |

25. When included in the speech input equipment of a broadcast station, what purpose do peak limiting amplifiers serve?
(a) They improve the fidelity of the audio signal.
(b) They reduce the gain of the amplifier when the amplitude, at some predetermined value, is exceeded so as to prevent overmodulation.
(c) They prevent adjacent-channel interference resulting from overmodulation.
(d) They increase the primary coverage area of the station.
(e) Both (b) and (c) are true.

(a) | | (b) | | (c) | | (d) | | (e) | |

26. Which of the following design features would not be used in a limiting amplifier used for broadcast FM stations?
(a) A fast attack time and a slow release time.
(b) The inclusion of a compressor circuit to compress the dynamic range.
(c) A very fast limiting circuit so that noticeable overmodulation will not occur, thus reducing adjacent channel interference.
(d) The inclusion of a threshold control to set the point of limiting.
(e) Circuitry that will provide limiting at 100 percent for negative peaks but that will allow modulation of the positive peaks up to 125 percent for greater coverage area.

(a) | | (b) | | (c) | | (d) | | (e) | |

27. An unbalanced to ground microphone cable has:
(a) one inner conductor and one shield.
(b) two inner conductors and two shields.
(c) three inner conductors and one shield.
(d) three conductors.
(e) four conductors.

(a) | | (b) | | (c) | | (d) | | (e) | |

28. A microphone that can pick up sound in a complete circle is said to be:
(a) bidirectional. (b) dynamic.
(c) omnidirectional. (d) stereo.
(e) unidirectional.

(a) | | (b) | | (c) | | (d) | | (e) | |

29. When authorized, which of the following provisions regarding the multiplexing of the aural carrier of a broadcast TV carrier is false?
(a) The multiplex channel may be used for telemetry and alerting signals from the transmitter site to the control point being remote controlled.
(b) The instantaneous frequency of the subcarrier used to modulate the aural carrier shall fall within the range of 53 to 68 kHz.
(c) The maximum modulation of the aural carrier by the subcarrier shall not exceed 10 percent of the maximum permissible degree of modulation.
(d) Multiplexing is limited to the use of a single subcarrier.
(e) The instantaneous frequency of the subcarrier used to modulate the aural carrier shall fall within the range of 20 to 50 kHz.

(a) | | (b) | | (c) | | (d) | | (e) | |

30. In Fig. 6, the collector potential is +8.0 V. If the transistor's β is 35 and the voltage drop across the emitter-base junction may be ignored, the value of R_b is:
(a) 520 kohms (b) 470 kohms
(c) 660 kohms (d) 380 kohms
(e) 590 kohms

(a) | | (b) | | (c) | | (d) | | (e) | |

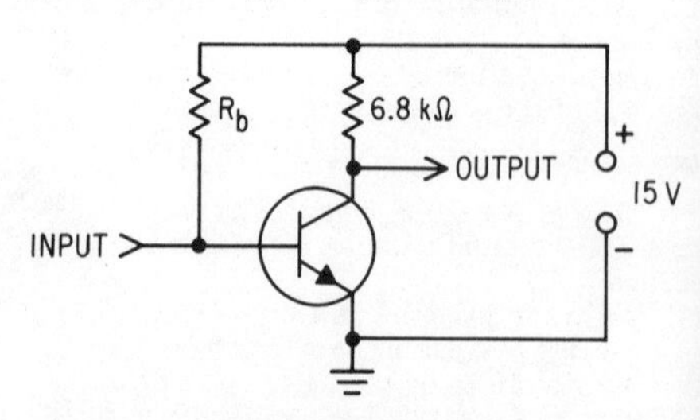

FIGURE 6

31. In the usual type of oscilloscope:
(a) sinusoidal horizontal deflection is used.
(b) Lissajous figures are used for vertical calibration.
(c) sawtooth horizontal deflection is used.
(d) magnetic deflection is employed.
(e) None of the above is true.

(a) | | (b) | | (c) | | (d) | | (e) | |

32. A system consisting of broadcast stations and interconnecting facilities which have been authorized by the Commission to operate in a controlled manner during a war, threat of war, state of public peril or disaster, or other national emergencies, is called the:
(a) Emergency Broadcast System.
(b) National Defense Emergency Authorization.
(c) Emergency Action Condition.
(d) Emergency Broadcast System Plan.
(e) National Defense Early Warning System.

(a) | | (b) | | (c) | | (d) | | (e) | |

33. A 6 VU line pad is often used between the output of a program amplifier and the telephone line to the transmitter to:
(a) reduce the phase distortion introduced by the phone line.
(b) improve the high frequency response of the line.
(c) reduce VU meter deviations due to the reflections from the telephone line.
(d) allow a higher program level to be transmitted over the telephone line.
(e) improve the low frequency response of the transmission system.

(a) | | (b) | | (c) | | (d) | | (e) | |

34. A lavalier microphone is used for:
(a) general pickups.
(b) mounting on a microphone boom.
(c) mounting on a desk stand.
(d) clipping on a person's clothing.
(e) mounting on floor stands.

(a) | | (b) | | (c) | | (d) | | (e) | |

35. Loading a parallel resonant circuit with a parallel resistance will:
(a) cause split tuning.
(b) increase the sharpness of the resonance curve.
(c) lower the resonant frequency of the tuned circuit.
(d) increase the frequency of resonance of the circuit.
(e) lower the Q of the tuned circuit.

(a) | | (b) | | (c) | | (d) | | (e) | |

36. The SCA used in conjunction with an FM broadcast station may not be authorized to transmit:
(a) program material of a broadcast nature for use at a later time.
(b) signals intended to delete main channel material.
(c) signals for remote cueing and order circuits.
(d) signals for telemetering functions.
(e) storecasting, special weather forecasting, and special time signals that have been subscribed to.

(a) | | (b) | | (c) | | (d) | | (e) | |

37. The waveform of the horizontal deflection voltage applied to an oscilloscope (employing electrostatic deflection) is a:
(a) sine wave. (b) square wave.
(c) sawtooth wave. (d) triangular wave.
(e) short duration pulse.

(a) | | (b) | | (c) | | (d) | | (e) | |

38. The field strength at a position five miles from a transmitter is 240 μV per m. If the transmitter power is tripled, what will be the new field intensity eight miles away from the transmitter?
(a) 150 μV per m (b) 450 μV per m
(c) 260 μV per m (d) 300 μV per m
(e) 275.8 μV per m

(a) | | (b) | | (c) | | (d) | | (e) | |

39. A crystal used in a transmitter's 3 MHz master oscillator has a negative temperature coefficient of 12 Hz per MHz per ^{o}C. At 30°C, the transmitter's output frequency is 24 MHz. If the temperature rises to 33°C, the radiated frequency will be:
(a) 24.000108 MHz (b) 23.999892 MHz
(c) 24.000864 MHz (d) 23.999136 MHz
(e) 23.999712 MHz

(a) | | (b) | | (c) | | (d) | | (e) | |

40. An RF amplifier is to be neutralized. The first step in performing this function is to:
(a) obtain a maximum input to the stage to be neutralized.
(b) disable the local oscillator.
(c) remove the plate voltage from the stage to be neutralized.
(d) set the neutralizing condenser for a maximum capacitance.
(e) tune all the stages of the transmitter to resonance.

(a) | | (b) | | (c) | | (d) | | (e) | |

41. The frame frequency of a TV broadcast station is:
(a) 10 per second (b) 20 per second
(c) 30 per second (d) 60 per second
(e) 120 per second

(a) | | (b) | | (c) | | (d) | | (e) | |

42. The operating power of a transmitter during daytime is 2500 W, but at nighttime the power is reduced by 1000 W. If the nighttime antenna current is 3 A, what is the daytime antenna current?
(a) 5 A (b) 7.5 A (c) 3.9 A (d) 4.75 A
(e) 2.32 A

(a) | | (b) | | (c) | | (d) | | (e) | |

43. The accelerator grid in the image section of the image-orthicon camera tube is used to:
(a) cause a fast trace and then a slow retrace of the electron beam scanning the picture area.
(b) accelerate the electrons emitted by the electron gun toward the target.
(c) move the electrons away from the photo cathode and toward the target.

(d) discharge the thin glass target in the scanning section of the tube.
(e) introduce a 180° phase shift so that the output video will not be inverted.

(a) | | (b) | | (c) | | (d) | | (e) | |

44. The circuit shown in Fig. 7 represents:
(a) a phase modulator circuit.
(b) a voltage degenerative feedback amplifier.
(c) a current degenerative feedback amplifier.
(d) an Armstrong frequency modulator.
(e) a Phasitron modulator.

(a) | | (b) | | (c) | | (d) | | (e) | |

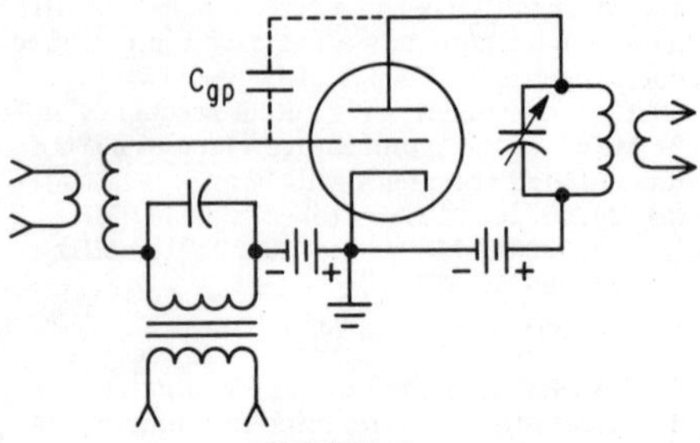

FIGURE 7

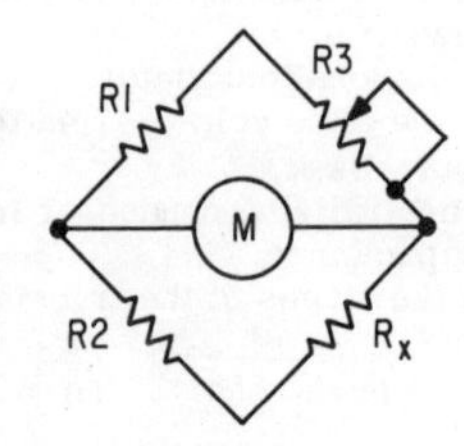

FIGURE 8

45. In Fig. 8, R1 = 100 ohms, R2 = 200 ohms, and R3 is adjusted for 300 ohms to produce a perfect balance of the bridge. What is the resistance of the unknown resistor, R_x?
(a) 15.0 ohms (b) 60.0 ohms
(c) 66.67 ohms (d) 150.0 ohms
(e) 600.0 ohms

(a) | | (b) | | (c) | | (d) | | (e) | |

46. Two coils share the same core, are wound in the same direction, and are connected in series. The total inductance is expressed by the formula:
(a) $L_T = (L_1 \times L_2)/(L_1 + L_2)$
(b) $L_T = L_1 + L_2 - 2M$
(c) $L_T = L_1 - L_2 - 2M$
(d) $L_T = L_1 + L_2 - 2k\sqrt{L_1 L_2}$
(e) $L_T = L_1 + L_2 + 2k\sqrt{L_1 L_2}$

(a) | | (b) | | (c) | | (d) | | (e) | |

47. An AM broadcast station operates on a carrier frequency of 1320 kHz and its antenna is 372 feet high. What percentage of a wavelength is the height of the antenna?
(a) 24.9% (b) 37.35% (c) 49.8%
(d) 74.7% (e) 99.6%

(a) | | (b) | | (c) | | (d) | | (e) | |

48. What is the total conductance of the circuit shown in Fig. 9?
(a) 2.5×10^{-4} mho (b) 5.0×10^{-4} mho
(c) 7.5×10^{-4} mho (d) 1.0×10^{-3} mho
(e) 1.25×10^{-3} mho

(a) | | (b) | | (c) | | (d) | | (e) | |

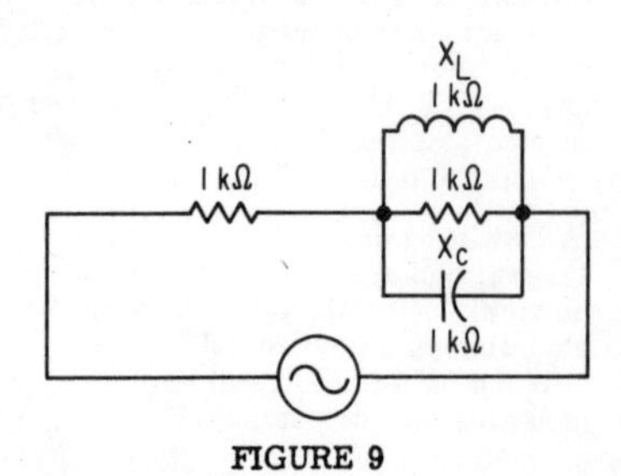

FIGURE 9

49. A field strength of 60 mV per m develops 2.5 V in a vertical antenna. What is the height in feet of the antenna?
(a) 41.6 (b) 68.3 (c) 98.4 (d) 136.7
(e) 273.4

(a) | | (b) | | (c) | | (d) | | (e) | |

50. The relationship between the power radiated from an antenna and the field strength is
(a) directly proportional to the power radiated.
(b) directly proportional to the square of the power radiated.
(c) proportional to the square root of the power radiated.
(d) directly proportional to the square root of the cube of the power radiated.
(e) There is no simple relationship between the field strength and the power radiated.

(a) | | (b) | | (c) | | (d) | | (e) | |

Element 4, Test 4

1. Where is a shotgun microphone used?
(a) For general pickups.
(b) On a boom for directional pickups.
(c) On a desk stand for announcements.
(d) Inside of musical instruments.
(e) On a floor stand for announcements.

(a) | | (b) | | (c) | | (d) | | (e) | |

2. The broadcast station program log contains the:
(a) time of the start of each program.
(b) time of the end of each program.
(c) name of each program.
(d) times of station identification.
(e) All the above are true.

(a) | | (b) | | (c) | | (d) | | (e) | |

3. Which of the basic design features listed below must be included in a television system to prevent frame bars from moving vertically across the screen?
(a) Interlaced scanning must be provided to synchronize the picture frame.
(b) Pre-emphasis must be used to improve the system's ability to hold sync.
(c) The field frequency must be equal to the commercial power supply frequency.
(d) To prevent frame bars from moving across the screen, extra large filter capacitors must be used.
(e) None of the above is true.

(a) | | (b) | | (c) | | (d) | | (e) | |

4. The percentage distribution required to produce white light in a color television picture is:
(a) 30% blue, 59% red, 11% green.
(b) 11% red, 30% blue, 59% green.
(c) 11% blue, 30% red, 59% green.
(d) 11% red, 30% green, 59% blue.
(e) 11% green, 30% red, 59% blue.

(a) | | (b) | | (c) | | (d) | | (e) | |

5. Where are overload relays normally used in broadcast transmitters?
(a) They are used in the feed line to the antenna to guard against transient overloads.
(b) They are placed in the final RF power supply to guard against filter capacitor failures.
(c) They are used in the plate circuits of some vacuum tubes to operate if the plate current exceeds the plate current rating of the tubes.
(d) They are used in the grid circuit of the RF amplifier to protect the tube against excess excitation.
(e) They are used in low-level modulator stages to protect the transmitter from overmodulation.

(a) | | (b) | | (c) | | (d) | | (e) | |

6. For TV broadcast stations, the effective radiated power of the aural transmitter is what percent of the peak radiated power of the visual transmitter?
(a) Not less than 5% nor more than 10%.
(b) Not less than 10% nor more than 20%.
(c) Not less than 10% nor more than 15%.
(d) Not less than 5% nor more than 25%.
(e) 25% $\pm$ 2%.

(a) | | (b) | | (c) | | (d) | | (e) | |

7. In the transmitter block diagram in Fig. 1, which stage could be removed without altering the output frequency?
(a) 1 (b) 2 (c) 3 (d) 4 (e) 5

(a) | | (b) | | (c) | | (d) | | (e) | |

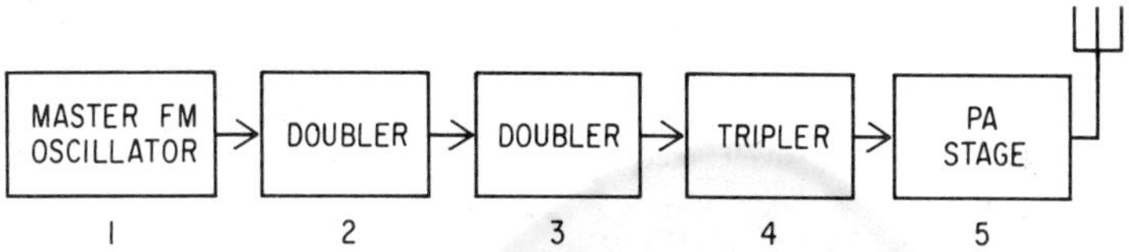

MASTER FM OSCILLATOR → DOUBLER → DOUBLER → TRIPLER → PA STAGE

1 2 3 4 5

FIGURE 1

8. In the transmitter block diagram of Fig. 1, what would be the effect on the output frequency of interchanging stages 3 and 4?
(a) The output frequency would be doubled.
(b) The output frequency would be tripled.
(c) The output frequency would be divided by three.
(d) The output frequency would be halved.
(e) No change.

(a) | | (b) | | (c) | | (d) | | (e) | |

9. At a station where the tower lights are not continuously monitored by an alarm device, how often should the lights be visually checked?
(a) Once each hour from sundown to sunup.
(b) Once each four-hour period from sundown to sunup.
(c) Once at sundown and at 12 midnight or at the end of the broadcast day.
(d) Once each 24-hour period.
(e) None of the above is true.

(a) | | (b) | | (c) | | (d) | | (e) | |

10. Refer to Fig. 2. E1 and E2 are connected to an ac signal source of 0.1 V. The output to ground E_O indicates a +6 V dc output and no signal voltage. What would be a possible cause?
(a) R1 is an open circuit.
(b) R2 is an open circuit.
(c) R3 is an open circuit.
(d) R5 is misadjusted.
(e) There is no -6 V supply voltage.

(a) | | (b) | | (c) | | (d) | | (e) | |

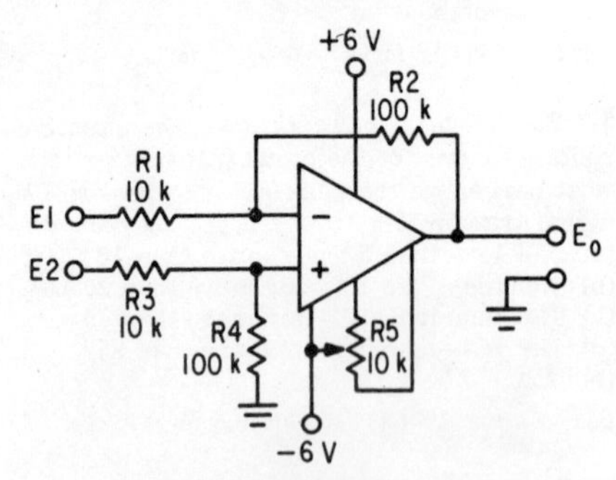

FIGURE 2

11. Two equal inductors are connected in series-opposing. If the coefficient of coupling is 0.5, the total inductance is:
(a) equal to the self-inductance of one coil.
(b) equal to twice the self-inductance of one coil.
(c) equal to half the self-inductance of one coil.
(d) equal to 1.5 times the self-inductance of one coil.
(e) equal to zero.

(a) | | (b) | | (c) | | (d) | | (e) | |

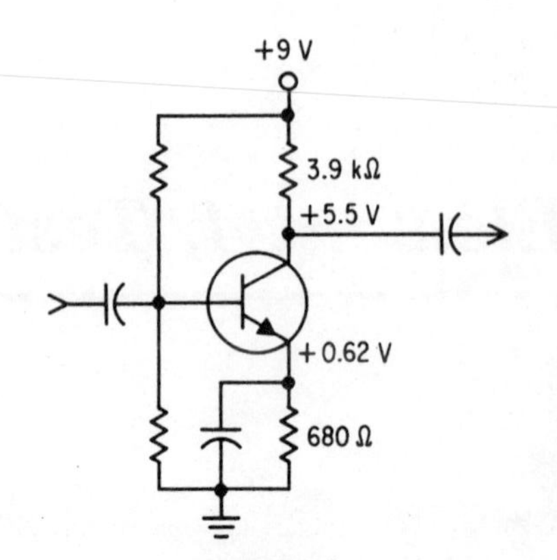

FIGURE 3

12. In Fig. 3, the value of the transistor's β is:
(a) 50 (b) 60 (c) 70 (d) 80 (e) 90

(a) | | (b) | | (c) | | (d) | | (e) | |

13. Which of the following is required to be entered in a station's operating log?
(a) The time and result of an auxiliary transmitter test.
(b) A notation each week of the calibration check of automatic recording devices.
(c) An entry whenever frequency measurements are made, including the date performed and description of method used.
(d) An entry whenever the output power meter is calibrated, as required by R&R 73.689.
(e) An entry noting each interruption of the carrier wave where restoration is not automatic, giving its cause and duration followed by the signature of the person restoring operation (if a licensed operator other than the licensed operator on duty).

(a) | | (b) | | (c) | | (d) | | (e) | |

14. An FM broadcast transmitter operating on 106.2 MHz uses a doubler, tripler, and quadrupler in its frequency multiplier stages. If the transmitter is 60 percent directly modulated by a 2 kHz tone, what is the frequency swing at the oscillator?
(a) ±2 kHz (b) ±45 kHz (c) ±3.875 kHz
(d) ±3.750 kHz (e) ±1.875 kHz

(a) | | (b) | | (c) | | (d) | | (e) | |

15. The lowest frequency to be passed through a common-emitter class A single-ended transistor stage is 20 Hz. The emitter resistance is 1 kohm. What value of capacitance should be used across the emitter resistance?
(a) 80 μF (b) 159 μF (c) 125 μF
(d) 100 μF (e) 8 μF

(a) | | (b) | | (c) | | (d) | | (e) | |

16. Two vertical antennas are placed in line with each other 90° apart, one to the north and one to the south. The current amplitudes fed to both antennas are equal in amplitude, with the current of the south antenna lagging the current of the north antenna by 90°. What will the horizontal radiation pattern be?
(a) Bidirectional, north and south.
(b) Bidirectional, east and west.
(c) Cloverleaf, with north, south, east, and west lobes.
(d) Unidirectional, with a lobe toward the north.
(e) Cardioid, with a lobe toward the south.

(a) | | (b) | | (c) | | (d) | | (e) | |

17. While monitoring the antenna current meter of a low-level plate-modulated AM transmitter, a "dip" in the antenna current is observed. Which of the following conditions would not be the cause of the condition?
(a) Insufficient excitation into the modulated RF amplifier.
(b) Insufficient bias at the modulated stage.
(c) 100 percent modulation being reached but not exceeded.
(d) Poor regulation of a common power supply.
(e) Excessive overloading of the modulated RF amplifier.

(a) | | (b) | | (c) | | (d) | | (e) | |

18. A VU meter is normally used across a program transmission line of which of the following characteristic impedances?
(a) 50 ohms (b) 72 ohms (c) 125 ohms
(d) 300 ohms (e) 600 ohms

(a) | | (b) | | (c) | | (d) | | (e) | |

19. For the best frequency stability, which of the following types of thermostatic control should be used for the crystal or oscillator oven?
(a) Bimetallic type, with hermetically sealed contact points.
(b) Bimetallic type, using a mercury tube as a switch.
(c) Mercury-thermometer type.
(d) Thermistor.
(e) A variable resistor in series with the heater element to control the heater current.

(a) | | (b) | | (c) | | (d) | | (e) | |

20. A low-frequency noise introduced by the turntable mechanics is known as:
(a) flutter. (b) rumble.
(c) wow. (d) pitch.
(e) rms.

(a) | | (b) | | (c) | | (d) | | (e) | |

21. When variable H or T pads becomes noisy or intermittent, the contact points and leaf springs may be cleaned by:
(a) using a clean, soft cloth and jeweler's rouge.
(b) using a burnishing tool.
(c) lightly filing with a point file.
(d) spraying with contact cleaner and wiping dry with a soft, clean cloth.
(e) using a crocus cloth.

(a) | | (b) | | (c) | | (d) | | (e) | |

22. The upper and lower limits of the carrier frequency of a standard broadcast station with an assigned carrier frequency of 1240 kHz are:
(a) 1, 239, 998 Hz to 1, 240, 002 Hz
(b) 1, 239, 380 Hz to 1, 240, 620 Hz
(c) 1, 233, 830 Hz to 1, 246, 200 Hz
(d) 1, 239, 980 Hz to 1, 240, 020 Hz
(e) 1, 239, 830 Hz to 1, 240, 200 Hz

(a) | | (b) | | (c) | | (d) | | (e) | |

23. If a licensed operator goes on duty at 6:00 A.M. and off duty at 2:00 P.M., how many times must he sign the operating log?
(a) 2 times (b) 8 times (c) 16 times
(d) 32 times (e) None of the above is true.

(a) | | (b) | | (c) | | (d) | | (e) | |

24. What is the hardest material used for a record player stylus?
(a) Ruby (b) Sapphire (c) Diamond
(d) Jade (e) Steel

(a) | | (b) | | (c) | | (d) | | (e) | |

25. A broadcast station antenna having a radiation resistance of 52 ohms is radiating 5 kW of RF power. If the antenna current is doubled, what power would the station be transmitting?
(a) 10, 000 W (b) 15, 000 W (c) 20, 000 W
(d) 25, 000 W (e) 30, 000 W

(a) | | (b) | | (c) | | (d) | | (e) | |

26. Which of the following types of emission is used in the sound transmitter of a TV broadcast station?
(a) A3 (b) A3J (c) A5 (d) F3 (e) F5

(a) | | (b) | | (c) | | (d) | | (e) | |

27. If the power delivered to an antenna is 2.75 kW when the antenna current is 6.7 A, what will be the increase in power if the current is raised to 9.6 A?
(a) 3.94 kW (b) 2.90 kW (c) 5.62 kW
(d) 1.19 kW (e) 2.19 kW

(a) | | (b) | | (c) | | (d) | | (e) | |

28. Which of the following statements is true of the design features included in the turnstile antenna used for broadcast TV antennas?
(a) The turnstile antenna achieves its wide band width in the same manner as the folded dipole antenna.
(b) The turnstile antenna achieves its gain by the use of reflectors.
(c) By notching in the sides of each batwing, additional current flows in the longer upper and lower sections, reducing the radiation in the vertical plane while improving the gain in the horizontal plane.

(d) The common impedance of the east-west and north-south elements of a three-bay antenna is 150 ohms.
(e) Each section is one wavelength high.

(a) | | (b) | | (c) | | (d) | | (e) | |

29. When an AM transmitter uses high level modulation, the modulating signal is added in the:
(a) phase correction network.
(b) buffer output circuit.
(c) oscillator output circuit.
(d) intermediate stage.
(e) final output stage.

(a) | | (b) | | (c) | | (d) | | (e) | |

30. When used in audio-limiting amplifiers, VU meters may be used to:
(a) indicate the amount of carrier shift introduced by the limiting amplifier, in VU.
(b) measure the output program level of the limiter.
(c) measure the incoming program level of the limiter.
(d) measure the amount of limiting caused by the limiter, in VU.
(e) (b), (c), and (d) are true.

(a) | | (b) | | (c) | | (d) | | (e) | |

31. A resonant tank circuit consists of a 150 μH coil, whose resistance is 5 ohms, in parallel with a 250 pf capacitor. What is the resonant frequency of the circuit?
(a) 821 kHz (b) 260 kHz (c) 2.6 MHz
(d) 725 kHz (e) 980 kHz

(a) | | (b) | | (c) | | (d) | | (e) | |

32. In the circuit of Question 31, what is the value of Q?
(a) 77.5 (b) 155 (c) 49 (d) 65.5
(e) 125

(a) | | (b) | | (c) | | (d) | | (e) | |

33. In the circuit of Question 31, what is the value of the impedance at resonance?
(a) 5 ohms (b) 775 ohms (c) 7.75 kohms
(d) 77.5 kohms (e) 120 kohms

(a) | | (b) | | (c) | | (d) | | (e) | |

34. In the circuit of Question 31, what is the value of the bandwidth?
(a) 2.65 kHz (b) 10.6 kHz (c) 155 kHz
(d) 5.3 kHz (e) 21.2 kHz

(a) | | (b) | | (c) | | (d) | | (e) | |

35. If a part of an AM broadcast station's speech equipment, where in the system are limiting amplifiers used?
(a) Following the preamplifiers and before the mixing line of the audio console so that the amplitude of the tape machines, turntables, and mike lines will be equal as fed to the mixer.
(b) They are used to drive telephone program lines so the peak amplitude of the program level will not cause crosstalk in other telephone lines.
(c) They are used in remote amplifiers to eliminate intensified peaks of the audio signal that would cause crosstalk in the telephone lines.
(d) At the transmitter location ahead of the AGC amplifier that feeds the audio input terminals of the transmitter.
(e) After the AGC amplifier and before the input terminals to the transmitter.

(a) | | (b) | | (c) | | (d) | | (e) | |

36. Referring to the operational amplifier circuit of Fig. 4, which of the following statements is true?
(a) The input impedance of the circuit is 100 kohms.
(b) The voltage gain $E_O/E1$ of the circuit is +2 for the frequencies in the passband.
(c) The voltage gain $E_O/E1$ of the circuit is -1 for frequencies in the passband.
(d) The voltage gain $E_O/E1$ of the circuit is +1 for frequencies in the passband.
(e) The input impedance of the circuit is 200 kohms.

(a) | | (b) | | (c) | | (d) | | (e) | |

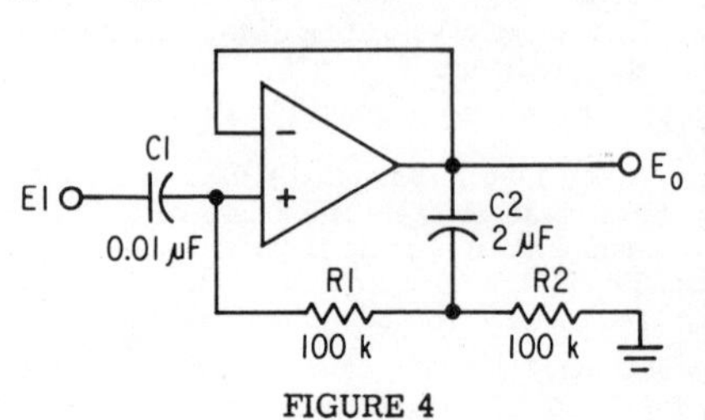

FIGURE 4

37. Which of the following is not a requirement of a diplexer used in broadcast TV antenna systems?
(a) It must include a vestigial sideband filter.
(b) It must have a minimum power absorption.
(c) It must provide very low crosstalk between the video and aural transmitters.
(d) It must provide a VSWR of less than 1.1.
(e) It must provide two outputs, one for the east-west and one for the north-south coaxials.

(a) | | (b) | | (c) | | (d) | | (e) | |

38. The visual carrier frequency tolerance of a television broadcast station is:
(a) 0.0005% (b) $\pm$500 Hz (c) $\pm$1000 Hz
(d) $\pm$1500 Hz (e) $\pm$2000 Hz

(a) | | (b) | | (c) | | (d) | | (e) | |

39. When a thermocouple ammeter is used in the transmission line of a standard broadcast antenna system, what methods may be

used to protect the ammeter from lightning?
(a) A shorting switch may be used across the ammeter when not in use.
(b) A series fuse may be used.
(c) The meter may be placed in the ground return lead of the antenna system.
(d) A large inductance may be placed in series with the meter.
(e) Lightning arrestors may be used in all cases.

(a) | | (b) | | (c) | | (d) | | (e) | |

40. In specifying a crystal unit, it is important to be aware of the type of oscillator circuit so that the:
(a) correct holder size can be ordered.
(b) crystal can be correlated to the oscillator.
(c) crystal oven can be properly calibrated.
(d) cost of the various components can be kept to a minimum.
(e) oven temperature can be calculated as a function of the ambient temperature.

(a) | | (b) | | (c) | | (d) | | (e) | |

41. A two-wire transmission line that is one and one-half wavelengths long is terminated by a resistive load of 125 ohms. The surge impedance of the line is 70 ohms. What is the input impedance to the line?
(a) 223 ohms (b) 125 ohms (c) 93.4 ohms
(d) 70 ohms (e) 39.2 ohms

(a) | | (b) | | (c) | | (d) | | (e) | |

42. The pre-emphasis circuit of an FM broadcast station transmitter employs a time constant of:
(a) 60 μs (b) 65 μs (c) 70 μs
(d) 75 μs (e) 80 μs

(a) | | (b) | | (c) | | (d) | | (e) | |

43. Why is impedance matching in broadcast speech input equipment important?
(a) It improves the energy transfer.
(b) It prevents overloading of the audio lines.
(c) It improves the fidelity.
(d) It prevents crosstalk between the various audio lines.
(e) Both (a) and (c) are true.

(a) | | (b) | | (c) | | (d) | | (e) | |

44. For an antenna current meter, what is the maximum permissible full-scale reading for square law scales?
(a) One and one-half the minimum normal reading.
(b) Twice the minimum normal reading.
(c) Three times the minimum normal reading.
(d) Four times the minimum normal reading.
(e) Five times the minimum normal reading.

(a) | | (b) | | (c) | | (d) | | (e) | |

45. What percentage of the peak carrier level is the blanking level in a monochrome TV signal?
(a) 12.5% (b) 25% (c) 50% (d) 65%
(e) 75%

(a) | | (b) | | (c) | | (d) | | (e) | |

46. For high-powered, high-level modulated AM transmitters, class B modulators are preferred over class A modulators because:
(a) the modulator stage can be operated in push-pull.
(b) class B operation produces less distortion.
(c) class B operation is more efficient than class A operation.
(d) the class B stage requires less driving power.
(e) a single-ended stage may be used instead of a push-pull stage, using two tubes.

(a) | | (b) | | (c) | | (d) | | (e) | |

47. If a resistance is inserted in series with a series resonant circuit, how would the Q and impedance be affected?
(a) The Q and the impedance would increase.
(b) The Q will increase and the impedance will decrease.
(c) The Q will decrease and the impedance will increase.
(d) The Q and the impedance will remain unchanged.
(e) The Q and the impedance would decrease.

(a) | | (b) | | (c) | | (d) | | (e) | |

48. An RF transmitter amplifier is neutralized to:
(a) reduce the possibility of harmful interference to other stations.
(b) prevent self-oscillation of an RF stage.
(c) prevent the generation of unwanted frequencies.
(d) protect the RF stage from excessive power dissipation.
(e) All the above are true.

(a) | | (b) | | (c) | | (d) | | (e) | |

49. An oscillator may be isolated from the following RF stages by:

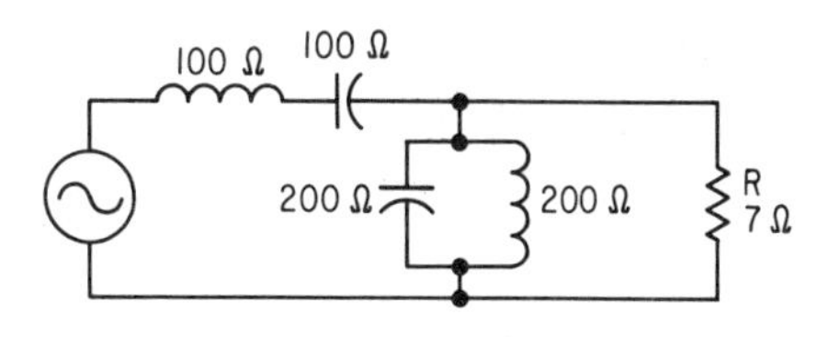

FIGURE 5

(a) a Faraday screen.
(b) a low-reluctance shield.
(c) an isolation transformer.
(d) a buffer amplifier.
(e) All of the above are true.

(a) | | (b) | | (c) | | (d) | | (e) | |

50. In Fig. 5, if a current of 2.8 A RMS flows through R, what must the value of the source voltage be?
(a) 560 V (b) 280 V (c) 19.6 V
(d) 28 V (e) 196 V

(a) | | (b) | | (c) | | (d) | | (e) | |

Element 4, Test 5

1. The term "crystal activity" is:
(a) the molecular activity of quartz with elevated temperature.
(b) a measure of the ability of quartz to oscillate after the oscillator power is cut off.
(c) limited to AT cuts only.
(d) a function of the equivalent resistance of a crystal, usually related to the oscillator load capacitance.
(e) an important consideration in designing for the smallest crystal size.

(a) | | (b) | | (c) | | (d) | | (e) | |

2. A decibel represents:
(a) an impedance ratio.
(b) the smallest amount of change in sound that the human ear can recognize.
(c) a VSWR.
(d) the amplitude of sound in units of VU.
(e) a change in velocity of sound waves.

(a) | | (b) | | (c) | | (d) | | (e) | |

3. With zero carrier level as a reference, what is the normal blanking pulse level used in composite television signals for commercial broadcast television in the United States?
(a) 12.5% (b) 25.0% (c) 63.5% (d) 75.0%
(e) 100%

(a) | | (b) | | (c) | | (d) | | (e) | |

4. Crystal drive level is defined as:
(a) the frequency at which the crystal oscillates.
(b) the output level of the crystal oscillator.
(c) the dc voltage applied across the crystal unit.
(d) the power dissipated in the crystal unit.
(e) another way of defining the capacity ratio of the crystal unit.

(a) | | (b) | | (c) | | (d) | | (e) | |

5. What is the total admittance of the circuit shown in Fig. 1?
(a) 2.5×10^{-4} mho (b) 7.07×10^{-4} mho
(c) 3.57×10^{-4} mho (d) 5.0×10^{-4} mho
(e) 5.0×10^{-3} mho

(a) | | (b) | | (c) | | (d) | | (e) | |

6. Electric field intensity when measured at a distance from the antenna is measured in:
(a) dB. (b) mV or μV per meter.
(c) W per second. (d) A per meter.
(e) Joules.

(a) | | (b) | | (c) | | (d) | | (e) | |

7. When combined with a 100 kohm resistor, what size capacitor would be required to produce a 25 μ second time constant?
(a) 250 pF (b) 0.250 μF (c) 400 μF
(d) 40 pF (e) 0.001 μF

(a) | | (b) | | (c) | | (d) | | (e) | |

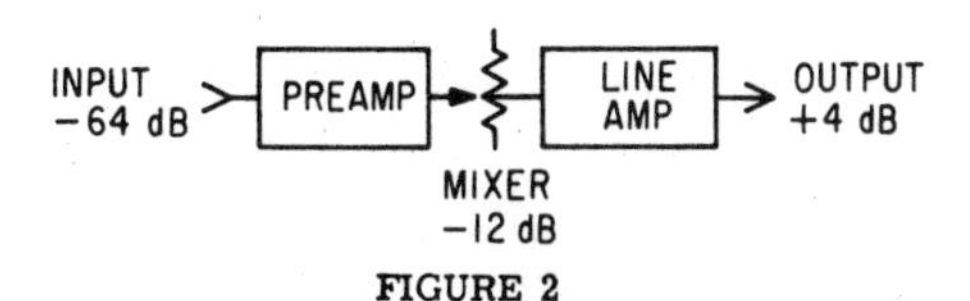

FIGURE 2

8. In the audio block diagram shown in Fig. 2, all inputs and outputs are 600 ohms. What must the gain of the preamp plus the line amplifier be to produce a +4 dB output?
(a) 100 (b) 3160 (c) 1000 (d) 10,000
(e) 31,600

(a) | | (b) | | (c) | | (d) | | (e) | |

9. Referring to the operational amplifier circuit of Fig. 3, which of the following statements is false?
(a) The gain of the circuit with respect to E1 is -10.

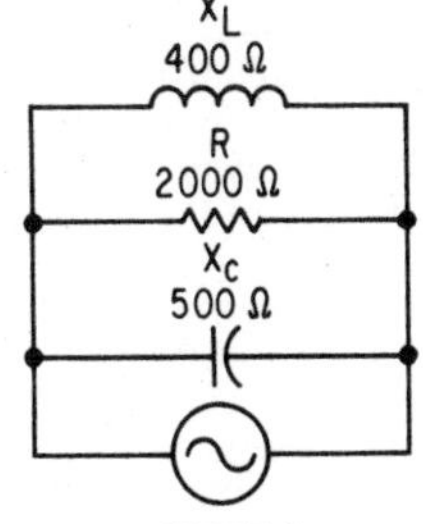

FIGURE 1

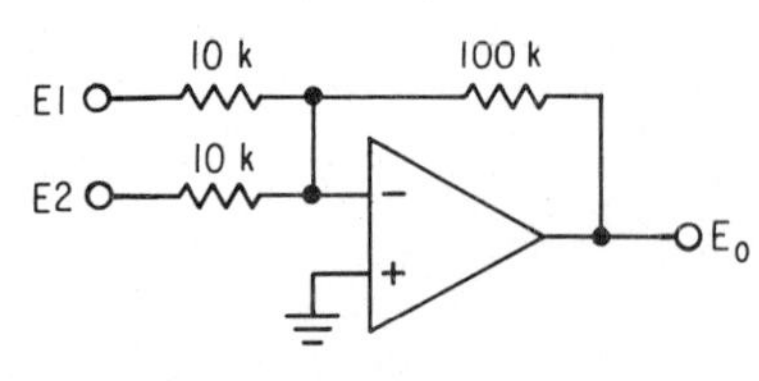

FIGURE 3

(b) The gain of the circuit with respect to E2 is -10.
(c) The signal E1 mixes with the signal E2.
(d) The signal E1 is applied in equal proportions with the signal E2 so that E_0 = -10(E1 + E2).
(e) The voltage at the inverting and noninverting inputs of the operational amplifier are virtually the same.

(a) | | (b) | | (c) | | (d) | | (e) | |

10. Modulation produced in an RF amplifier stage preceding the final RF stage is called:
(a) high-level modulation.
(b) pre-emphasis modulation.
(c) low-level modulation.
(d) the Armstrong system of modulation.
(e) None of the above is true.

(a) | | (b) | | (c) | | (d) | | (e) | |

11. The RF power input to an antenna is 15,000 W, but its effective radiated power is 60 kW. The antenna power gain is:
(a) 12 dB (b) 6 dB (c) 4 dB (d) 3 dB
(e) 8 dB

(a) | | (b) | | (c) | | (d) | | (e) | |

12. What type of microphone requires a power source?
(a) The dynamic microphone.
(b) The carbon microphone.
(c) The crystal microphone.
(d) The lavalier microphone.
(e) The ribbon microphone.

(a) | | (b) | | (c) | | (d) | | (e) | |

13. A microphone that employs a very light strip of corrugated aluminum suspended within a permanent magnetic field is called the:
(a) velocity microphone.
(b) dynamic microphone.
(c) ribbon microphone.
(d) condenser microphone.
(e) Both (a) and (c) are true.

(a) | | (b) | | (c) | | (d) | | (e) | |

14. In Fig. 4, the transistor's β is 40. If the emitter-base junction voltage drop is 0.3 V, the base current is:
(a) 33 μA (b) 131 μA (c) 31 μA
(d) 133 μA (e) 2.44 μA

(a) | | (b) | | (c) | | (d) | | (e) | |

15. In Fig. 4, the collector potential is:

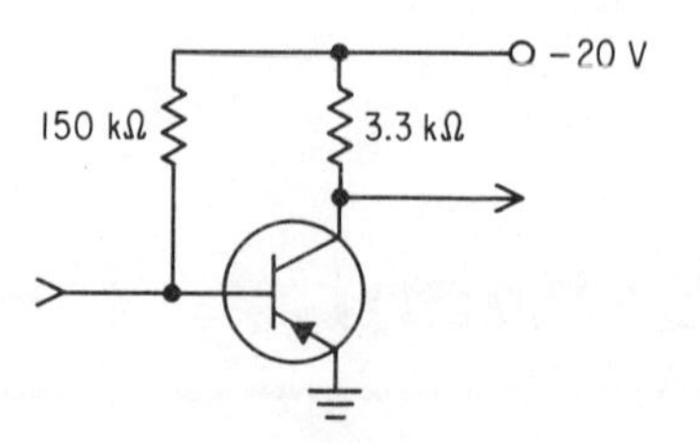

FIGURE 4

(a) +17.3 V (b) -2.7 V (c) -20 V
(d) -12.7 V (e) -17.3 V

(a) | | (b) | | (c) | | (d) | | (e) | |

16. The frequency tolerance of a noncommercial FM broadcast station licensed for more than 10 W is:
(a) ±1 kHz (b) ±2 kHz (c) ±3 kHz
(d) ±0.003% (e) ±0.005%

(a) | | (b) | | (c) | | (d) | | (e) | |

17. Which of the following items regarding the operation of the antenna-tower lighting should be included in the station's maintenance log?
(a) The time the tower lights are turned on and off each day, if manually controlled.
(b) The time the daily check on proper operation of the tower lights was made, if an automatic alarm system is not provided.
(c) Any observed or otherwise known failure of the tower lights.
(d) Date and time the failure of the lights was observed, or otherwise noted.
(e) All the above are true.

(a) | | (b) | | (c) | | (d) | | (e) | |

18. If the gain of an AF amplifier without feedback is 20, and the gain with feedback is 15, what is the feedback percentage?
(a) 0.16% (b) 1.7% (c) 5.0% (d) 10.0%
(e) 20.0%

(a) | | (b) | | (c) | | (d) | | (e) | |

19. What is a microphone pad used for?
(a) Increasing the microphone output.
(b) Decreasing the microphone output.
(c) Filtering out hum.
(d) Changing microphone patterns.
(e) Recording in stereo.

(a) | | (b) | | (c) | | (d) | | (e) | |

20. Which of the following cannot be performed by an H or T pad composed of pure resistance?
(a) DC isolation between two circuits.
(b) Impedance matching between two circuits.
(c) A variable attenuation.
(d) A fixed attenuation.
(e) A constant input and output impedance.

(a) | | (b) | | (c) | | (d) | | (e) | |

21. The operating power of a standard FM broadcast station is determined by:
(a) (Antenna current)^2 x Antenna resistance.
(b) Plate voltage x Plate current x Efficiency factor F.
(c) Antenna power input x Antenna power gain.
(d) Power output from the final stage minus the transmission line loss.
(e) (RF voltage measured at the antenna feed point)^2 x Antenna resistance.

(a) | | (b) | | (c) | | (d) | | (e) | |

22. At what point in the antenna system must the antenna current be measured for a standard broadcast station?
(a) At a point of maximum current.
(b) At a point of maximum voltage.
(c) Between the antenna tuning inductance and the tuning condenser.
(d) At the antenna terminal of the transmitter.
(e) At the point where the antenna resistance was determined.

(a) | | (b) | | (c) | | (d) | | (e) | |

23. The frequency and tolerance of a color TV broadcast station chrominance-subcarrier is designated as:
(a) 3.850000 MHz $\pm$ 1%
(b) 3.579545 MHz $\mp$ 10 Hz
(c) 3.580000 MHz $\pm$ 0.05%
(d) 3.579545 MHz $\mp$ 5 Hz
(e) 3.579545 MHz $\pm$ 0.01%

(a) | | (b) | | (c) | | (d) | | (e) | |

24. In the relationship between the field strength of a radiated wave and the distance from the transmitter's antenna, the field strength is:
(a) directly proportional to the square of the distance.
(b) inversely proportional to the square of the distance.
(c) directly proportional to the square root of the distance.
(d) inversely proportional to the distance.
(e) directly proportional to the distance.

(a) | | (b) | | (c) | | (d) | | (e) | |

25. In an FM broadcast receiver, how is the audio signal detected?
(a) By detecting the amount of frequency swing of the carrier, which produces an output amplitude that corresponds to the amplitude of the original modulating signal.
(b) The frequency produced in the output of an FM detector is determined by the rate of frequency deviation of the carrier.
(c) The amplitude variations of the carrier determine the amplitude of the output signal from the detector.
(d) Both (a) and (b) are true.
(e) Both (b) and (c) are true.

(a) | | (b) | | (c) | | (d) | | (e) | |

26. The input voltage to a matched 500-ohm transmission line is 1250 V RMS. The transmission line current at the antenna feed point is 2.435 A RMS. What is the power loss on the transmission line?
(a) 250 W (b) 218 W (c) 173 W
(d) 160 W (e) 104 W

(a) | | (b) | | (c) | | (d) | | (e) | |

27. An FM transmitter has a frequency deviation of 20 kHz and transmits an audio range of 200 Hz to 8 kHz. If the bandwidth allowed is 50 kHz, what is the value of the deviation ratio?
(a) 100 (b) 3.0 (c) 40.0 (d) 2.5
(e) 7.5

(a) | | (b) | | (c) | | (d) | | (e) | |

28. A two-wire transmission line, which is 3/4 λ long, is terminated by a resistive load of 75 ohms. The input impedance to the line is measured and found to be 27 ohms. What is the surge impedance of the line?
(a) 48 ohms (b) 102 ohms (c) 51 ohms
(d) 45 ohms (e) 33.7 ohms

(a) | | (b) | | (c) | | (d) | | (e) | |

29. In Fig. 5, E2 is at ground potential while an 0.1 V RMS signal is applied between E1 and ground. The E_0 output, as measured on an oscilloscope, reads 1.55 V ac peak. What if anything is a possible cause?
(a) R5 is an open circuit.
(b) R2 is an open circuit.
(c) R3 is an open circuit.
(d) The output is normal.
(e) R4 is an open circuit.

(a) | | (b) | | (c) | | (d) | | (e) | |

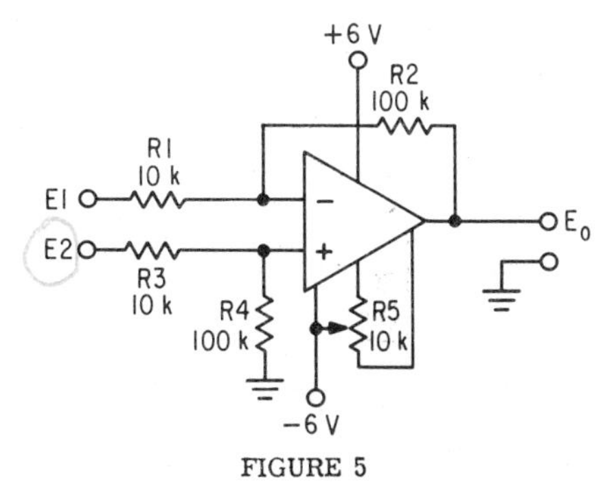

FIGURE 5

30. Of the following statements, which is true with regard to the frequency range of the "I" amplifier output and the phase of the chrominance subcarrier feed to the "I" balanced modulator?
(a) From 2.08 to 5.08 MHz, the subcarrier is in phase with the "Q" subcarrier.
(b) From 3.08 to 4.08 MHz, the subcarrier is shifted 90° with respect to the "Q" subcarrier.

(c) From 2.08 to 5.08 MHz, the subcarrier is referred to as the "in-phase subcarrier."
(d) From 3.08 to 4.08 MHz, the subcarrier is referred to as the "quadrature subcarrier."
(e) From 3.08 to 4.08 MHz, the subcarrier is referred to as the "in-phase subcarrier."

(a) | | (b) | | (c) | | (d) | | (e) | |

31. When a PN junction is forward biased, the:
(a) junction resistance is decreased.
(b) junction resistance is increased.
(c) junction resistance remains the same.
(d) electrons flow from the N material to the P material.
(e) Both (a) and (d) are correct.

(a) | | (b) | | (c) | | (d) | | (e) | |

32. What percentage of the peak carrier level, as represented by the synchronizing pulse, is the reference white level required by the FCC?
(a) 12.5% (b) 25% (c) 50% (d) 70%
(e) 80%

(a) | | (b) | | (c) | | (d) | | (e) | |

33. In audio amplifiers, variable T or H pads provide:
(a) good equalization.
(b) control of the amplitude of the signal.
(c) good isolation between circuits.
(d) impedance matching.
(e) Both (b) and (d) are true.

(a) | | (b) | | (c) | | (d) | | (e) | |

34. It is desired to pick up the conversation of a number of people in a round-table discussion. What type of microphone and placement would be best suited for this application?
(a) A bidirectional velocity microphone, placed in the center of the table.
(b) An omnidirectional dynamic microphone, placed at the center of the table.
(c) A unidirectional microphone placed at the edge of the table.
(d) A cardioid microphone, placed in the center of the table.
(e) A unidirectional microphone, placed at the center of the table.

(a) | | (b) | | (c) | | (d) | | (e) | |

35. Time delays are used to apply high voltage to the anodes of mercury vapor rectifier tubes some time after the application of the filament voltage, for the purpose of:
(a) preventing damage to the cathode.
(b) protecting the filter condensers against the inverse peak voltage.
(c) reducing the peak surge current through the filter chokes.
(d) insuring the safety of operating personnel.
(e) none of the above.

(a) | | (b) | | (c) | | (d) | | (e) | |

36. An overload relay is generally a:
(a) voltage-sensitive relay.
(b) current-sensitive relay.
(c) power-sensitive relay.
(d) volt-amp-reactive (VAR) sensitive relay.
(e) None of the above are true.

(a) | | (b) | | (c) | | (d) | | (e) | |

37. The plate supply voltage of a class C RF power amplifier is 750 V, and the mean dc level of plate current is 1.3 A. The plate efficiency of the amplifier is 72 percent. What is the plate dissipation?
(a) 878 W (b) 705 W (c) 273 W
(d) 178 W (e) 93.6 W

(a) | | (b) | | (c) | | (d) | | (e) | |

38. In Fig. 5, there is a -6 V dc signal between E_O and ground. The inputs E1 and E2 have a 0.05 V ac input signal applied between them. What is a possible cause?
(a) R5 is not adjusted properly.
(b) R1 is an open circuit.
(c) R4 is an open circuit.
(d) There is no + 6 V supply voltage.
(e) R3 is an open circuit.

(a) | | (b) | | (c) | | (d) | | (e) | |

39. For maximum stability, how should a crystal oscillator be tuned?
(a) Tune for maximum oscillator current output.
(b) Tune for minimum oscillator plate current.
(c) Tune to the point of minimum plate current; then reduce the tank circuit capacity.
(d) Tune for maximum oscillator plate current; then add slightly more tank circuit capacity.
(e) Tune to the point of minimum plate current; then add slightly more tank circuit capacity.

(a) | | (b) | | (c) | | (d) | | (e) | |

40. Two vertical broadcast station antennas are spaced 135° apart and are fed equal amplitude currents in-phase. What horizontal radiation pattern would be produced by the antennas?
(a) A triangular pattern, consisting of three main lobes, one of them in the direction of a line bisecting the two antennas.
(b) A bidirectional pattern, approximately oblong, with the two lobes at right angles to a line bisecting the two antennas.
(c) A bidirectional pattern, approximately oblong, with the two lobes in the direction of a line bisecting the two towers.
(d) A cardioid pattern, at right angles to a line bisecting the two antennas.
(e) A cloverleaf pattern, with lobes in the direction of a line bisecting the two antennas and at right angles to it.

(a) | | (b) | | (c) | | (d) | | (e) | |

41. What is the frequency tolerance of an FM broadcast station?
(a) $\pm$2 Hz (b) $\pm$20 Hz (c) $\pm$200 Hz
(d) $\pm$2000 Hz (e) $\pm$2500 Hz

(a) | | (b) | | (c) | | (d) | | (e) | |

42. The aspect ratio of the picture width to the picture height as transmitted by commercial television broadcast stations is:
(a) 4 to 3 (b) 5 to 4 (c) 3 to 2
(d) 6 to 4
(e) as specified in the individual station license.

(a) | | (b) | | (c) | | (d) | | (e) | |

43. Why are grounded grid amplifiers sometimes used in the VHF and UHF frequencies?
(a) They require much less driving power.
(b) They can be water-cooled, whereas grounded cathode tubes cannot.
(c) A triode tube can be used without neutralization circuits.
(d) Larger tubes can be used in the grounded grid configuration.
(e) No tank circuits are required in the RF class C grounded grid amplifiers at VHF and UHF frequencies.

(a) | | (b) | | (c) | | (d) | | (e) | |

44. What is the "0 VU" reference level used by broadcast studios and telephone companies?
(a) 1 W in 500 ohms
(b) 1 mW in 600 ohms
(c) 10 mW in 500 ohms
(d) 1 mW in 250 ohms
(e) 10 mW in 250 ohms

(a) | | (b) | | (c) | | (d) | | (e) | |

45. To convert a power ratio in VU to one in dB:
(a) divide the value in VU by 0.707.
(b) divide the value in VU by 1.414.
(c) no conversion is necessary.
(d) multiply the value in VU by 0.9.
(e) divide the value in VU by 0.9.

(a) | | (b) | | (c) | | (d) | | (e) | |

46. If the power input to a 70-ohm transmission line is 10.08 kW, what is the current flowing through the line?
(a) 6 A (b) 12 A (c) 36 A (d) 144 A
(e) 288 A

(a) | | (b) | | (c) | | (d) | | (e) | |

47. The field frequency of a monochrome TV broadcast station is:
(a) 10 per sec (b) 20 per sec
(c) 30 per sec (d) 60 per sec
(e) 120 per sec

(a) | | (b) | | (c) | | (d) | | (e) | |

48. In a transmitter employing low-level grid modulation, which of the following items could cause variations in the final amplifier plate current?
(a) Excessive resistance in the grid bias power supply.
(b) Insufficient operating bias on the grid of the modulated RF amplifier.
(c) Insufficient loading of the plate circuit of the modulated RF amplifier.
(d) Excessive RF excitation to the grid of the modulated RF amplifier.
(e) All of the above are true.

(a) | | (b) | | (c) | | (d) | | (e) | |

49. When comparing the amplitude of the carrier wave to the amplitude of the harmonic content, the unit of measurement that is generally used is:
(a) volts per meter. (b) the watt.
(c) the decibel. (d) the joule.
(e) the coulomb.

(a) | | (b) | | (c) | | (d) | | (e) | |

50. The main purpose of including a limiter as part of standard broadcast speech equipment is to:
(a) compress the audio amplitude range of the program signal.
(b) improve the dynamic range of the program signal.
(c) limit the audio frequency response to the audio range.
(d) keep high amplitude peaks of the program signal from overmodulating the transmitter, which would cause frequency splatter.
(e) eliminate the need for manual control at the studio.

(a) | | (b) | | (c) | | (d) | | (e) | |

Element 4, Test 6

1. In Fig. 1, the transistor's β is 100. Ignoring the voltage drop across the emitter-base junction and the voltage drop across R_{B1} due to the flow of base current, the approximate value of the emitter current is:
(a) 0.51 mA (b) 0.53 mA (c) 0.55 mA
(d) 0.59 mA (e) 0.62 mA

(a) | | (b) | | (c) | | (d) | | (e) | |

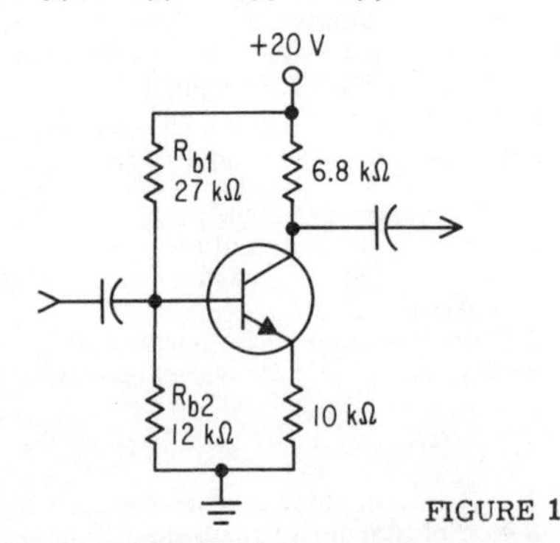

FIGURE 1

2. In Fig. 1, the collector/emitter voltage is:
(a) 15.8 V (b) 6.2 V (c) 9.6 V
(d) 7.8 V (e) 8.0 V

(a) | | (b) | | (c) | | (d) | | (e) | |

3. Which of the following statements is not true regarding the characteristics of an AGC amplifier used in broadcast speech equipment?
(a) They include circuitry to reduce the dynamic range of the audio signals.
(b) They reduce the necessity of "riding the gain" of the program level.
(c) They help maintain a higher percentage of modulation than manual control.
(d) They help maintain a constant program level when switching between different pickup facilities.
(e) They maintain a constant output level by "clipping off the peaks" of the audio signal at a predetermined level.

(a) | | (b) | | (c) | | (d) | | (e) | |

4. For standard broadcast stations, the indirect method of determining the RF power feed to the antenna includes the use of:
(a) the square of the antenna current multiplied by the radiation resistance of the antenna.
(b) the transmission line voltage multiplied by the transmission line current.
(c) the RF final plate voltage multiplied by the final plate current.
(d) the product of the antenna current and the final plate voltage.
(e) the product of the final plate current, plate voltage, and efficiency factor F.

(a) | | (b) | | (c) | | (d) | | (e) | |

5. For a standard broadcast station, which of the following is not a requirement for the indicating instruments used in the last radio stage of the transmitter?
(a) Length of the scale shall be not less than 2 3/10 inches.
(b) Accuracy shall be at least 2 percent of the full-scale reading.
(c) The maximum rating of the meter shall be such that it does not read off scale during modulation.
(d) Scale shall have at least 50 divisions.
(e) Full-scale reading shall not be greater than five times the minimum normal indication.

(a) | | (b) | | (c) | | (d) | | (e) | |

6. What is the total current flowing in the circuit shown in Fig. 2?
(a) 10.22 A (b) 12.09 A (c) 16.14 A
(d) 18.25 A (e) 21.54 A

(a) | | (b) | | (c) | | (d) | | (e) | |

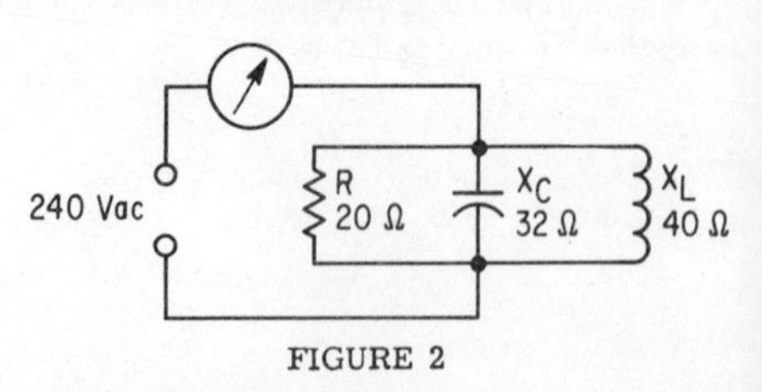

FIGURE 2

7. The most obvious reason for avoiding extremely small values of crystal load capacitance is:
(a) the higher cost of small load capacitors.
(b) the frequency instability caused by small capacitance changes, as well as the influence they have on the impedance of the crystal.
(c) that it increases the temperature coefficient of the crystal.

(d) that the use of other than the standard C_L of 30 pF or 32 pF is not economical.
(e) the inconvenience of getting enough capacitance range to achieve a desired frequency adjustment.

(a) | | (b) | | (c) | | (d) | | (e) | |

8. The principal reason for providing a crystal oven is to:
(a) maintain better stability by avoiding the effect of ambient temperature changes.
(b) insure maximum thermo-isolation from other components.
(c) improve the power consumption of the crystal.
(d) insure minimum frequency shift.
(e) Both (a) and (d) are true.

(a) | | (b) | | (c) | | (d) | | (e) | |

9. Of the following signals, which one is not transmitted as part of the standard television video signal?
(a) Pilot subcarrier.
(b) Blanking pulse.
(c) Video RF carrier.
(d) Vertical sync pulses.
(e) Horizontal sync pulses.

(a) | | (b) | | (c) | | (d) | | (e) | |

10. Reflectometers are used in television stations to:
(a) enhance the brilliance of the video signal.
(b) measure the incident and reflected voltage on the transmission line and to indicate the ratio.
(c) increase the brilliance of the audio signal.
(d) prevent overmodulation.
(e) None of the above is true.

(a) | | (b) | | (c) | | (d) | | (e) | |

11. Which of the following statements regarding hollow pressurized coaxial transmission lines is false?
(a) Their surge or characteristic impedance is generally within the range of 200 to 600 ohms.
(b) For a given size, they have higher breakdown voltage ratings.
(c) They have less loss per foot when used at 200 MHz.
(d) Less moisture will accumulate on the inside of the coaxial, increasing its efficiency.
(e) None of the above is true.

(a) | | (b) | | (c) | | (d) | | (e) | |

12. In a television broadcast station, the video carrier frequency is separated from the aural carrier frequency by:
(a) 6 MHz (b) 4.9 MHz (c) 4.7 MHz
(d) 4.5 MHz (e) 4.0 MHz

(a) | | (b) | | (c) | | (d) | | (e) | |

13. The impedance of a standard AM broadcast antenna may be measured by:
(a) a Wheatstone bridge. (b) a VTVM.
(c) an impedance bridge.
(d) a decade resistance box.
(e) a tuning stub with the aid of the Smith chart.

(a) | | (b) | | (c) | | (d) | | (e) | |

14. Which of the following is not the name of a microphone pattern?
(a) Dynamic. (b) Omnidirectional.
(c) Semidirectional. (d) Cardioid.
(e) Bidirectional.

(a) | | (b) | | (c) | | (d) | | (e) | |

15. An AM transmitter is high level modulated by a single audio tone. If the modulation percentage is 80 percent, what is the percentage of total sideband power in relation to the unmodulated carrier power?
(a) 24% (b) 32% (c) 40% (d) 20%
(e) 64%

(a) | | (b) | | (c) | | (d) | | (e) | |

16. The output of a microphone is rated in:
(a) volts. (b) mV. (c) mA. (d) amps.
(e) dBm.

(a) | | (b) | | (c) | | (d) | | (e) | |

17. When used in broadcast station transmitters, recycling relays will normally operate when:
(a) transient overloads exist within their circuit.
(b) the final amplifier plate voltage is too low.
(c) the plate voltage is 15 percent above its normal value.
(d) the transmitter is undermodulated.
(e) the carrier of the transmitter drifts slightly off frequency.

(a) | | (b) | | (c) | | (d) | | (e) | |

18. In the transmission and reproduction of broadcast color television, the "Y" signal has a bandwidth of:
(a) 0 to 4.2 MHz and supplies the luminance (or brightness) signal for black and white (monochrome) reproduction.
(b) 0 to 0.5 MHz and contains the chrominance signals for the cyan to orange color range.
(c) 0 to 0.5 MHz and contains the chrominance signals for the yellow-green to purple color range.
(d) 0 to 1.5 MHz and contains the chrominance signals for the yellow-green to purple color range.
(e) 0 to 1.5 MHz and is the luminance (or brightness) signal used for fine detail in color transmissions and for the black and white reproduction in monochrome transmission.

(a) | | (b) | | (c) | | (d) | | (e) | |

19. Cathode disintegration in a mercury vapor rectifier tube may not be caused by:
(a) exceeding the rated peak inverse voltage.
(b) excessive plate current.
(c) an excessive filament warm-up period.
(d) using the tube with a capacitor input filter.
(e) a simultaneous application of the plate and filament voltages.

(a) | | (b) | | (c) | | (d) | | (e) | |

20. The "Emergency Broadcast Condition" is:
(a) a special identification signal given when a national emergency exists.
(b) the time period between the emergency notification and the emergency termination.
(c) the National Defense Emergency Authorization.
(d) a grave national crisis.
(e) None of the above is true.

(a) | | (b) | | (c) | | (d) | | (e) | |

21. An audio frequency power amplifier has an output voltage of 55 V as measured across an external load resistor of 500 ohms. What is the power in the load resistor?
(a) 4.25 W (b) 6.05 W (c) 6.80 W
(d) 9.10 W (e) 10.90 W

(a) | | (b) | | (c) | | (d) | | (e) | |

22. When two or more microphones are used together, they must be:
(a) placed facing each other.
(b) placed facing away from each other.
(c) connected together by means of a matching transformer.
(d) correctly phased.
(e) connected to each other through a line equalizer.

(a) | | (b) | | (c) | | (d) | | (e) | |

23. A microphone that depends for its operation on the emf produced by a moving coil of wire within a magnetic field is the:
(a) condenser microphone.
(b) ribbon microphone.
(c) carbon microphone.
(d) dynamic microphone.
(e) velocity microphone.

(a) | | (b) | | (c) | | (d) | | (e) | |

24. The speed of a turntable may be checked by using a:
(a) tuning fork.
(b) dynamometer.
(c) stroboscopic disc.
(d) velometer.
(e) deviation meter.

(a) | | (b) | | (c) | | (d) | | (e) | |

25. Two vertical broadcast antennas are in line with each other, one to the north and the other to the south, spaced 135° apart. Equal currents are fed to both antennas, with the current fed to the south antenna lagging in phase by 135°. The horizontal radiation pattern would be:
(a) basically circular, with half-nodes appearing in the north and the south directions.
(b) basically circular, with half-nodes appearing in the east and west directions.
(c) two cardioid patterns toward the east and west.
(d) a bidirectional pattern east and west, with a complete null between the two antennas.
(e) one large main lobe toward the south and a minor lobe toward the north.

(a) | | (b) | | (c) | | (d) | | (e) | |

26. How often should the broadcast station antenna tower be painted?
(a) Once each year.
(b) Once each two years.
(c) Once each three years.
(d) Once each five years.
(e) It should be cleaned or repainted as often as necessary to maintain good visibility.

(a) | | (b) | | (c) | | (d) | | (e) | |

27. What is the standard tape width used for most consumer and semiprofessional reel-to-reel tape recorders?
(a) 1/8 in. (b) 1/4 in. (c) 1/2 in.
(d) 1 in. (e) 2 in.

(a) | | (b) | | (c) | | (d) | | (e) | |

28. What type of circuit is shown in the schematic in Fig. 3?
(a) A linear class C RF amplifier, with a class B modulator.
(b) A push-push class C RF amplifier, modulated by a push-push class B modulator.
(c) Phase-modulated FM transmitter.
(d) Low-level amplitude-modulated class C RF final amplifier, modulated with a push-pull class B amplifier.
(e) High-level amplitude-modulated class C RF amplifier, with a class B modulator.

(a) | | (b) | | (c) | | (d) | | (e) | |

29. For minimum distortion, the secondary impedance of T1 in Fig. 3 should be:
(a) 1 kohm (b) 2 kohms (c) 2.5 kohms
(d) 5 kohms (e) 10 kohms

(a) | | (b) | | (c) | | (d) | | (e) | |

30. In Fig. 3, what is the purpose of capacitor C5?
(a) It provides neutralization for the stage.
(b) It is an audio bypass capacitor.
(c) It provides negative voltage feedback for the RF stage to prevent self-oscillation.
(d) It is an RF bypass capacitor.
(e) None of the above is true.

(a) | | (b) | | (c) | | (d) | | (e) | |

31. During 100 percent modulation, the audio signal appearing across the secondary of T1

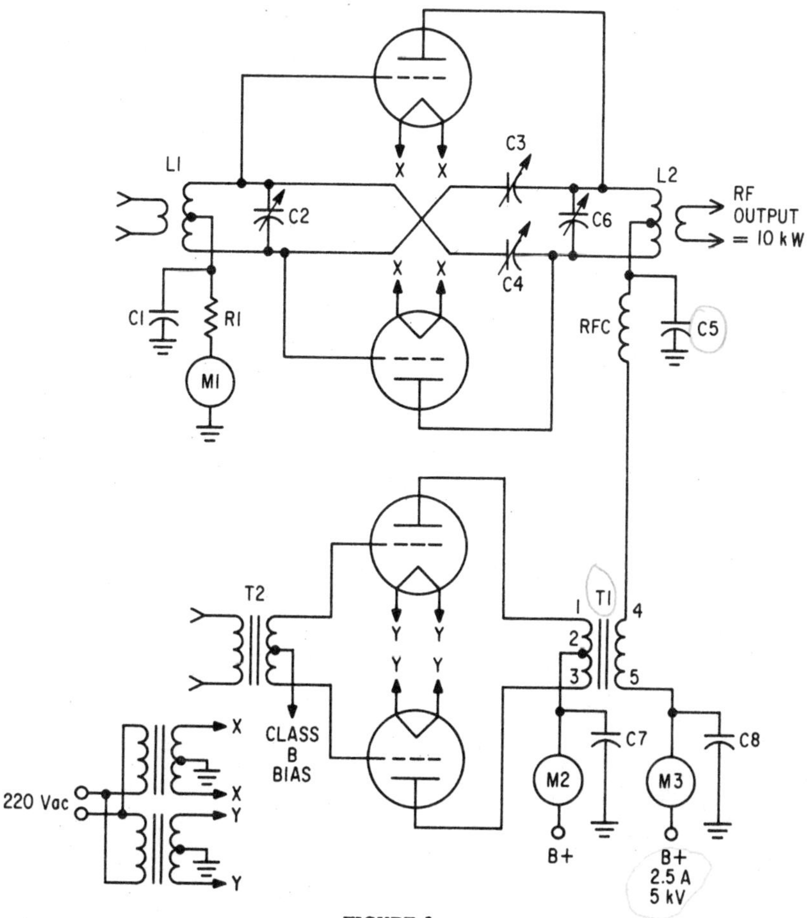

FIGURE 3

in Fig. 3 is positive at terminal 4 and negative at terminal 5. During modulation, the plate voltage with respect to ground would:
(a) double. (b) fall to zero.
(c) not change. (d) increase 50 percent.
(e) decrease 50 percent.

(a) | | (b) | | (c) | | (d) | | (e) | |

32. In Fig. 3, what is the plate efficiency of the RF modulated stage?
(a) 90% (b) 85% (c) 80% (d) 75%
(e) 70%

(a) | | (b) | | (c) | | (d) | | (e) | |

33. In Fig. 3, if the efficiency of the modulator stage is 65 percent, what is the required dc input power to the modulator stage for 100 percent modulation?
(a) 2.50 kW (b) 5.00 kW (c) 6.25 kW
(d) 9.62 kW (e) 10.00 kW

(a) | | (b) | | (c) | | (d) | | (e) | |

34. Tape is pulled across tape heads at a constant speed by:
(a) a take-up reel.
(b) a supply reel.
(c) a capstan and pinch roller.
(d) tape guides.
(e) pressure pads.

(a) | | (b) | | (c) | | (d) | | (e) | |

35. If an existing commercial TV broadcast station is allowed to add a commercial FM broadcast station and the latter is to be multiplexed with the TV antenna system, what parts of the system must be replaced?
(a) The vestigial sideband filter.
(b) The balun unit in the video transmitter.
(c) The diplexer.
(d) The 90^0 phase shift network used in the east-west bridge elements.
(e) None of the above are true.

(a) | | (b) | | (c) | | (d) | | (e) | |

36. An FM transmitter is 100 percent modulated by 1 kHz sine wave modulation. The transmission line current, without modulation, is 10 A. What is the transmission line current during modulation?
(a) 10.00 A (b) 11.75 A (c) 12.25 A
(d) 18.75 A (e) 22.50 A

(a) | | (b) | | (c) | | (d) | | (e) | |

37. The band of frequencies allotted to the FM broadcast services includes:
(a) 535 to 1605 kHz (b) 54 to 890 MHz
(c) 30 to 300 MHz (d) 88 to 108 MHz
(e) 300 to 3000 MHz

(a) | | (b) | | (c) | | (d) | | (e) | |

38. In Fig. 4, the open loop gain of the operational amplifier shown is 700,000. What is the actual gain of the circuit within its bandpass, in VU?
(a) +30 VU (b) +20 VU (c) Zero VU
(d) +100 VU (e) +40 VU

(a) | | (b) | | (c) | | (d) | | (e) | |

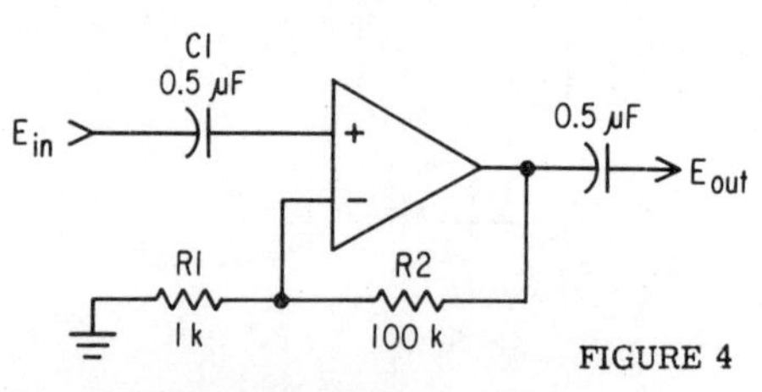

FIGURE 4

39. Radio frequency currents within the broadcast band are most commonly measured with:
(a) a hot wire ammeter.
(b) a thermocouple type ammeter.
(c) a vacuum tube ammeter.
(d) an iron vane ammeter.
(e) a D'Arsonval ammeter with a copper-oxide rectifier.

(a) | | (b) | | (c) | | (d) | | (e) | |

40. Two transistors have a tank circuit at the input that is connected to the bases in push-pull. The collectors are connected in parallel and have a tank circuit for the output. The ratio of the input to output frequency is:
(a) 1 to 1 (b) 1 to 2 (c) 2 to 1
(d) 1 to 3 (e) 3 to 1

(a) | | (b) | | (c) | | (d) | | (e) | |

41. Preamplifiers are commonly used in all commercial broadcast stations. The purpose of such amplifiers is to:
(a) provide an improved signal-to-noise ratio for the signal which is fed to the program mixer.
(b) increase the average modulation regardless of the type of program material.
(c) prevent overmodulation on high peaks.
(d) maintain the modulation at a constant value.
(e) provide the correct pre-emphasis.

(a) | | (b) | | (c) | | (d) | | (e) | |

42. Referring to the operational amplifier circuit of Fig. 5, which of the following statements is false?
(a) The low-frequency gain Eo/E1 of the circuit is -1.
(b) The circuit exhibits a high-frequency audio pre-emphasis of 75 microseconds time constant.
(c) The circuit exhibits a high-frequency audio de-emphasis of 10 microseconds time constant.
(d) The circuit would not be useful in providing audio high-frequency pre-emphasis in an FM receiver application.
(e) The circuit would be useful in providing audio high-frequency pre-emphasis in an FM receiver application.

(a) | | (b) | | (c) | | (d) | | (e) | |

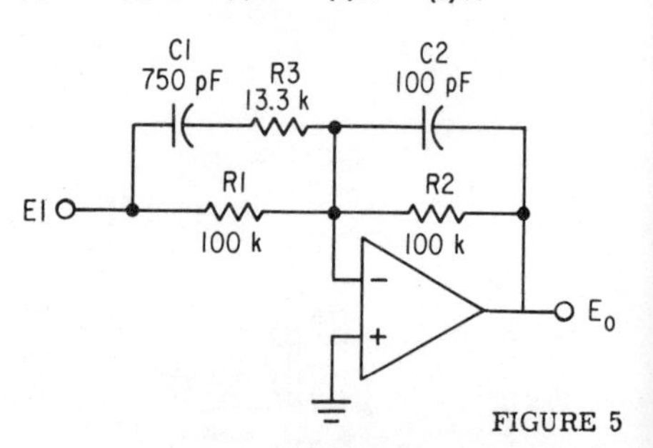

FIGURE 5

43. When a six-bay stacked ringed dipole array is used for an FM broadcast station antenna, it has a power gain of approximately:
(a) 1.6 (b) 3 (c) 5 (d) 6
(e) 10 or more

(a) | | (b) | | (c) | | (d) | | (e) | |

44. In a standard broadcast station, the degree of modulation that must generally be maintained should not be less than what percentage on peaks of frequent recurrence?
(a) 50% (b) 75% (c) 85% (d) 90%
(e) 100%

(a) | | (b) | | (c) | | (d) | | (e) | |

45. How may the inverse peak voltage be calculated for a full-wave power supply?
(a) The peak inverse voltage is equal to the peak-to-peak voltage of the secondary.
(b) Multiply the RMS secondary voltage of the power transformer by 1.414.
(c) Multiply the peak ac voltage of the secondary by 1.414.
(d) The peak inverse voltage is equal to half the peak voltage of the power transformer secondary.
(e) The peak inverse voltage is equal to the RMS value of the entire secondary wind-

ing times 1.414, less the drop in the conducting tube and the dc drop in the half of the transformer that is conducting.

(a) | | (b) | | (c) | | (d) | | (e) | |

46. A counterpoise system for use with an antenna is required when:
(a) a Hertz dipole is positioned near the ground.
(b) a dipole is neither vertical nor horizontal and produces an inclined plane of polarization.
(c) the ground is of poor conductive material, such as in a desert. The counterpoise system is then used to improve the antenna efficiency.
(d) the harmonic radiation is excessive and exceeds the FCC regulations.
(e) a Marconi antenna is operated remote from ground. The counterpoise system is then used to provide an apparent ground.

(a) | | (b) | | (c) | | (d) | | (e) | |

47. A nondirectional 50 kW broadcast station has an antenna current of 10 A. If the power were reduced to 5 kW, what would the new value of the antenna current be?
(a) 3.16 A (b) 7.07 A (c) 1.00 A
(d) 3.12 A (e) 5.00 A

(a) | | (b) | | (c) | | (d) | | (e) | |

48. Grid-leak bias is often used in class C RF power amplifiers because:
(a) it permits the output power to remain constant regardless of variations of input signal.
(b) it limits the plate current in the event the input signal is lost.
(c) it provides more linear operation.
(d) it makes possible a lower level of driving signal.
(e) it increases the value of the average plate current.

(a) | | (b) | | (c) | | (d) | | (e) | |

49. To attain good frequency stability in a crystal controlled oscillator:
(a) a low-temperature coefficient crystal should be employed.
(b) an X-cut crystal should be employed.
(c) a Y-cut crystal should be employed.
(d) a temperature-controlled crystal oven should be employed.
(e) Both (a) and (d) are true.

(a) | | (b) | | (c) | | (d) | | (e) | |

50. The circuit shown in Fig. 6 represents:
(a) audio tone control network.
(b) de-emphasis circuit.
(c) pre-emphasis circuit.
(d) audio-correction network for converting PM to FM.
(e) impedance matching network for connecting the microphone to the pre-amplifier.

(a) | | (b) | | (c) | | (d) | | (e) | |

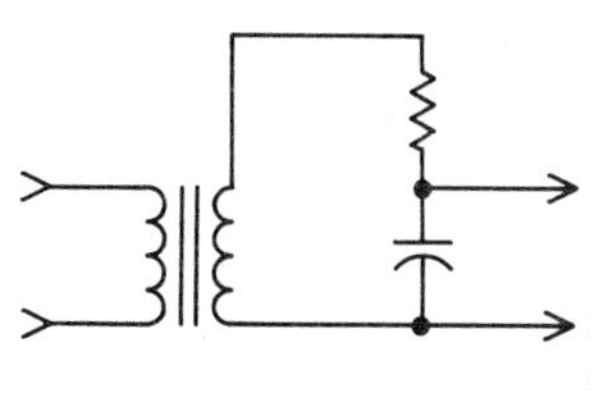

FIGURE 6

Element 4, Test 7

1. In referring to load capacitance, the reference is to:
(a) the ability of the crystal to operate with a 1 megohm or larger resistor in series with it.
(b) a measure of the crystal inductance.
(c) the frequency stability as determined by the power supply stability.
(d) the capacitive reactance appearing at the crystal terminals of the oscillator at the operating frequency.
(e) how thick the electrodes can be on the quartz plate without reducing the crystal activity.

(a) | | (b) | | (c) | | (d) | | (e) | |

2. In Fig. 1, what percentage of the peak carrier amplitude is indicated at point "M," and what is point "M" called?
(a) 12.5%; reference white level.
(b) 10%; blanking level.
(c) 12.5%; reference black level.
(d) 10%; picture level.
(e) 12.5%; horizontal blanking level.

(a) | | (b) | | (c) | | (d) | | (e) | |

3. The time represented by "N" in Fig. 1 is:
(a) 262.5 μs (b) 63.5 μs (c) 525.0 μs
(d) 10.16 μs (e) 15.750 μs

(a) | | (b) | | (c) | | (d) | | (e) | |

4. In Fig. 1, the blanking level is indicated by point:
(a) "L" (b) "K" (c) "M" (d) "J"
(e) "A"

(a) | | (b) | | (c) | | (d) | | (e) | |

5. In Fig. 1, the vertical blanking interval is indicated by letter:
(a) "B" (b) "D" (c) "E" (d) "F"
(e) "N"

(a) | | (b) | | (c) | | (d) | | (e) | |

6. In Fig. 1, the time interval indicated by point "D" is the:
(a) equalizing pulse interval.
(b) vertical sync pulse interval.
(c) vertical blanking interval.
(d) horizontal sync pulse interval.
(e) horizontal pulse width.

(a) | | (b) | | (c) | | (d) | | (e) | |

7. In Fig. 1, if a chrominance subcarrier were to be added to the figure, it would appear at point:
(a) "A" (b) "B" (c) "H" (d) "C"
(e) "I"

(a) | | (b) | | (c) | | (d) | | (e) | |

8. Crystal stability tolerance is defined as:
(a) resistance change with time.
(b) power dissipation with temperature variations.
(c) maximum permissible frequency change under stated conditions of temperature change, time, or other environmental changes.
(d) the flatness of the quartz surface as a result of dimensioning.
(e) the shunt capacity of the crystal unit.

(a) | | (b) | | (c) | | (d) | | (e) | |

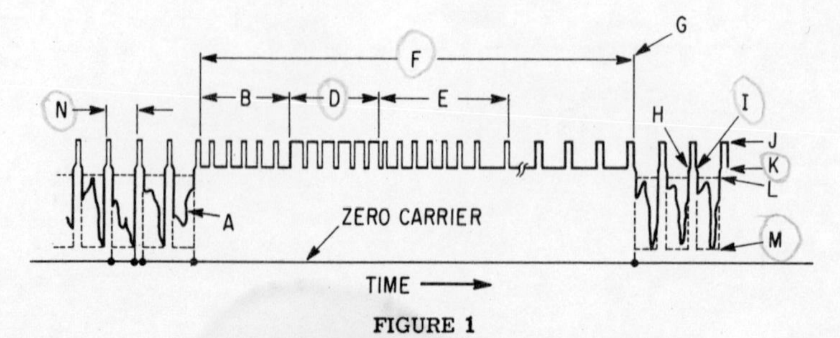

FIGURE 1

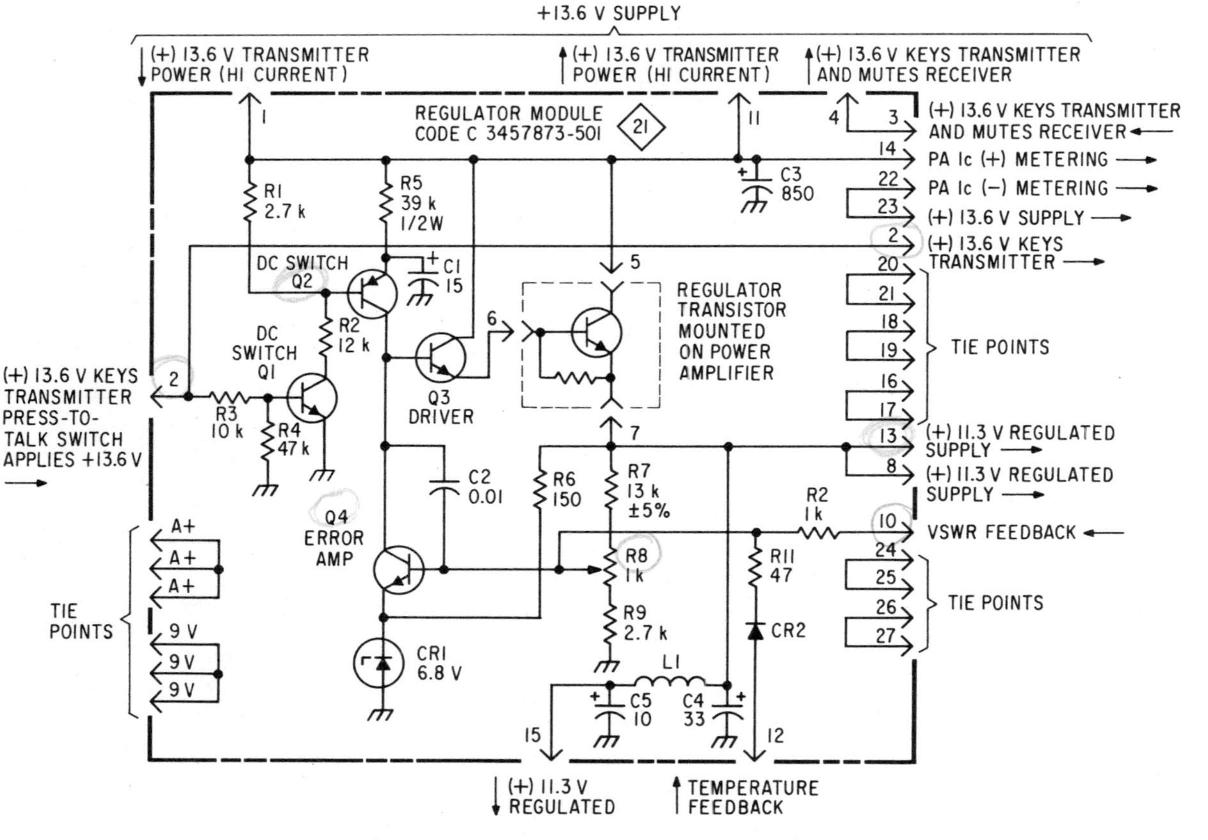

FIGURE 2 (Courtesy, RCA Corporation)

9. In the operation of transistor amplifiers:
(a) the forward bias of a class "A" amplifier is less than the forward bias of a class "B" amplifier.
(b) the forward bias on a class "B" amplifier is less than the forward bias of a class "C" amplifier.
(c) The forward bias on a class "A" amplifier is greater than the forward bias of a class "B" amplifier.
(d) the forward bias on a class "C" amplifier is greater than the forward bias on a class "B" amplifier.
(e) Both (a) and (b) are true.

(a) | | (b) | | (c) | | (d) | | (e) | |

10. RCA's regulator module (Fig. 2) is used in their vehicle radio systems for supplying regulated power from a nonregulated source. If potentiometer R8's variable slider is moved from the center position towards the top, the:
(a) collector of Q4 will become less positive.
(b) current will decrease through the regulator transistor.
(c) output voltage at point 14 will increase.
(d) collector current of Q3 will increase.
(e) Both (a) and (b) are true.

(a) | | (b) | | (c) | | (d) | | (e) | |

11. In Fig. 2, the output voltage at point 13 is found to be 12.2 V. A possible cause would be that:
(a) C1 is a short circuit.
(b) the base-emitter resistor of the regulator transistor is shorted.
(c) R8 is improperly set.
(d) Q4 has excessive leakage between the emitter-collector.
(e) Both (c) and (d) are true.

(a) | | (b) | | (c) | | (d) | | (e) | |

12. In Fig. 2, if + 13.6 V is not applied to input pin 2, this will cause:
(a) Q2 to conduct.
(b) Q1 to conduct.
(c) Q2 to be biased off.
(d) the regulator to conduct.
(e) an output at pin 8.

(a) | | (b) | | (c) | | (d) | | (e) | |

13. In Fig. 2, if the input voltage at point 10 causes the base of Q4 to become more positive, the:
(a) voltage at point 13 will decrease.
(b) voltage at point 13 will increase.
(c) collector-emitter current of Q3 will increase.
(d) voltage across CR1 will increase.
(e) voltage drop across the regulator transistor will decrease.

(a) | | (b) | | (c) | | (d) | | (e) | |

14. In Fig. 2, Q1 and Q2 act as dc switches to turn the power supply on when the "press to talk" button is operated. If + 13.6 V is then at input point 2:
(a) there will be no output at point 15.
(b) Q2 will not conduct.
(c) the regulator transistor will not conduct.
(d) Q2 will conduct.
(e) the voltage drop across R5 decreases.

(a) | | (b) | | (c) | | (d) | | (e) | |

15. Two microphones placed next to each other must be:
(a) on a "Y" cord.
(b) in phase with each other.
(c) of the same type.
(d) of different patterns.
(e) used only for stereo.

(a) | | (b) | | (c) | | (d) | | (e) | |

16. In broadcast TV color transmission, which of the following signals must be delayed about 1 microsecond to match the delay of other signals?
(a) The color burst signal.
(b) The "Q" signal.
(c) The "I" signal.

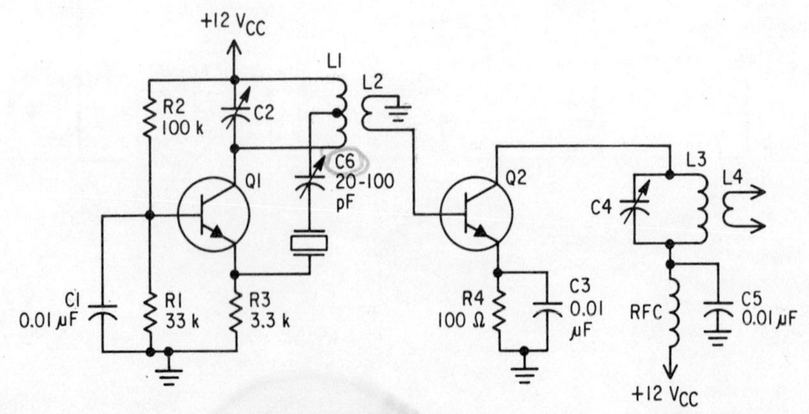

FIGURE 3

(d) The "Y" signal.
(e) The chrominance subcarrier signal.

(a) | | (b) | | (c) | | (d) | | (e) | |

17. Figure 3 is a typical oscillator and frequency multiplier stage that might be used in a TV transmitter. The crystal is operated at its third overtone and with a fundamental frequency of 9.3 MHz, approximately. The frequency multiplier is a tripler stage. If C6 were shorted:
(a) oscillations would cease.
(b) Q1 would go into saturation.
(c) the forward bias of Q2 would increase.
(d) the crystal drive level would increase and destroy the crystal.
(e) the oscillator would continue to operate but the output frequency would change slightly.

(a) | | (b) | | (c) | | (d) | | (e) | |

18. Which of the following best describes the operation of the ratio detector?
(a) It is a detector for frequency modulated waves whose output is an inverse ratio to two applied IF amplitudes at its input.
(b) It is an FM detector whose output is proportional only to the ratio of the input IF voltages and not to their amplitudes.
(c) It is a detector for AM waves, giving an AF output amplitude that is proportional to the IF input amplitude.
(d) It is a detector for frequency modulated waves whose AF output voltage amplitude varies directly with the input voltage amplitudes.
(e) It is a detector for frequency modulated waves that operates on the "slope detection" principle.

(a) | | (b) | | (c) | | (d) | | (e) | |

19. If the antenna input power is multiplied by the power gain of the antenna, the product is called the:
(a) authorized carrier power.
(b) effective radiated power.
(c) output power.
(d) peak power. (e) peak-to-peak power.

(a) | | (b) | | (c) | | (d) | | (e) | |

20. An audio transformer has a step-down ratio of 25 to 1. What is its impedance ratio for matching purposes?
(a) 25 to 1 (b) 5 to 1 (c) 1 to 25
(d) 1 to 5 (e) 625 to 1

(a) | | (b) | | (c) | | (d) | | (e) | |

21. Which of the following is considered to be the normal audio frequency band?
(a) 20 to 20,000 Hz (b) 10 to 15,000 Hz
(c) 30 to 30,000 Hz (d) 30 to 3,500 Hz
(e) 10 to 10,000 Hz

(a) | | (b) | | (c) | | (d) | | (e) | |

22. Two vertical broadcast antennas are spaced 90° apart, and the current in antenna B with respect to the current in antenna A is lagging in phase by 180°. If the currents in both antennas are equal in amplitude, the horizontal radiation pattern will be:
(a) bidirectional in a plane along a line bisecting the two towers.
(b) circular.
(c) at 45° to a line bisecting the two antennas.
(d) bidirectional and at right angles to the line bisecting the two antennas.
(e) cardioid with the main lobe in the direction of antenna A.

(a) | | (b) | | (c) | | (d) | | (e) | |

23. In a series resonant circuit, the resistance, the inductive reactance, and the capacitive reactance are each 80 ohms. If the frequency is reduced to 0.707 of its original value, what will be the impedance of the circuit?
(a) 80 ohms (b) 136 ohms (c) 160 ohms
(d) 98 ohms (e) 89.7 ohms

(a) | | (b) | | (c) | | (d) | | (e) | |

24. An ac series circuit consists of a 5 ohm resistance, 10 ohms of capacitive reactance, and 20 ohms of inductive reactance, with a current of 1.5 A in the circuit. What is the voltage across the series components?
(a) 52.5 V (b) 45.0 V (c) 30.0 V
(d) 16.8 V (e) 7.50 V

(a) | | (b) | | (c) | | (d) | | (e) | |

25. Which of the following is not a standard tape speed?
(a) 1 7/8 ips (b) 3 3/4 ips (c) 7 1/2 ips
(d) 33 1/3 ips (e) 15 ips

(a) | | (b) | | (c) | | (d) | | (e) | |

26. A klystron employs which of the following in its operation?
(a) Phase modulation.
(b) Frequency modulation (direct).
(c) Velocity modulation.
(d) Pulse modulation.
(e) Amplitude modulation, single sideband.

(a) | | (b) | | (c) | | (d) | | (e) | |

27. The field pattern of a directional two-element broadcast antenna will change if:
(a) changes occur in the common-point line current.
(b) the ratio of currents fed to the two driven elements changes.
(c) the phase between the two driven elements is kept at nominal.
(d) the transmitter output power is increased.
(e) the transmitter output power is decreased.

(a) | | (b) | | (c) | | (d) | | (e) | |

28. The frequency tolerance of a standard broadcast station operating at 640 kHz is:

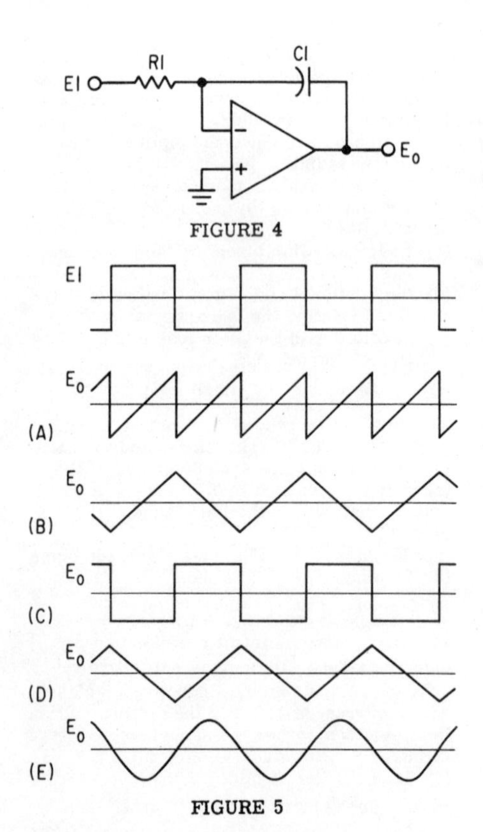

FIGURE 4

FIGURE 5

(a) ±0.0005% (b) ±0.0025% (c) ±2 kHz
(d) ± 20 Hz (e) ±2 Hz

(a) | | (b) | | (c) | | (d) | | (e) | |

29. Refer to the operational amplifier circuit of Fig. 4. Which of the output signal waveforms, E_O, in Fig. 5 corresponds to the square wave input signal E1?

(a) | | (b) | | (c) | | (d) | | (e) | |

30. When making dynamic (or amplitude) curves of a tape recorder, which of the following methods may be used?
(a) Using an audio sine-wave generator connected to the input of the amplifier and holding the amplitude constant while varying the frequency, plot the output amplitude variations in VU as a function of frequency.
(b) Using an audio sine wave generator connected to the input of the tape recorder and a VU meter properly terminating the output of the recorder, vary the frequency and record the input amplitude changes in VU to maintain a constant amplitude output.
(c) Using a properly terminated audio sweep generator connected to the input of the tape recorder, terminate the output of the recorder in a proper load, and with an ocilloscope triggered from the sweep generator record the variations in amplitude in VU as a function of frequency.
(d) Only (a) and (b) are true.
(e) (a), (b) and (c) are true.

(a) | | (b) | | (c) | | (d) | | (e) | |

31. Which of the following statements would be true of the characteristics of an AGC amplifier as used in broadcast speech equipment?
(a) They provide higher gain for lower level signals.
(b) They provide less gain for high level signals.
(c) The output of the AGC amplifier is virtually constant with widely varying inputs.
(d) AGC amplifiers reduce the dynamic range (amplitude) of the program signal.
(e) All of the above are true.

(a) | | (b) | | (c) | | (d) | | (e) | |

32. It is sometimes necessary to operate crystals in a temperature controlled oven. This is necessary to:
(a) decrease hysteresis losses.
(b) stabilize the feedback amplitude.
(c) increase the frequency stability.
(d) minimize the need for neutralization.
(e) Both (c) and (d) are true.

(a) | | (b) | | (c) | | (d) | | (e) | |

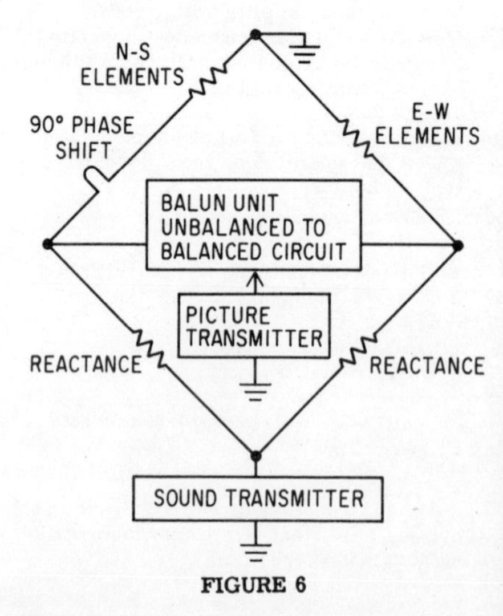

FIGURE 6

33. Figure 6 represents which of the following?
(a) A bridge-type diplexer.
(b) A resistance Wheatstone bridge.
(c) An antenna system for remote TV pickups.
(d) A duplexer antenna system.
(e) A balanced modulator.

(a) | | (b) | | (c) | | (d) | | (e) | |

34. A 50 kW FM transmitter is modulated by an 8000 Hz audio tone, which produces a frequency deviation of 20 kHz. What is the total RF power output, including the carrier and the sidebands?
(a) 50 kW (b) 62.5 kW (c) 71.6 kW
(d) 57.8 kW
(e) Cannot be determined from the information given.

(a) | | (b) | | (c) | | (d) | | (e) | |

35. In an AM broadcast directional antenna array operating on 925 kHz, two of the towers are separated by 120°. What is the actual distance in feet between the towers?
(a) 354.6 ft (b) 532 ft (c) 178 ft
(d) 266 ft (e) 712 ft

(a) | | (b) | | (c) | | (d) | | (e) | |

36. In the circuit shown in Fig. 7, what is the total impedance?
(a) 135 ohms (b) 96 ohms (c) 15 ohms
(d) 83.5 ohms (e) 106.7 ohms

(a) | | (b) | | (c) | | (d) | | (e) | |

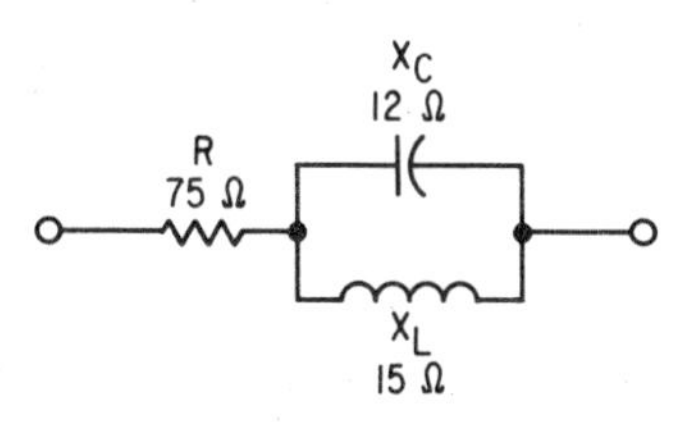

FIGURE 7

37. An X-cut crystal that has a temperature coefficient of - 12 Hz per MHz per degree centigrade has been calibrated at 45°C to 1.5 MHz. If the temperature increases to 50°C, at what frequency will the crystal oscillate?
(a) 1,499,910 Hz (b) 1,500,090 Hz
(c) 1,499,940 Hz (d) 1,500,060 Hz
(e) 1,499,100 Hz

(a) | | (b) | | (c) | | (d) | | (e) | |

38. A resonant tank circuit contains a capacitance of 187 pF and is resonant at 975 kHz. The inductance and resistance of the coil have not been measured. If the capacitor is removed and replaced by another whose capacitance is 294 pF, what will be the new resonant frequency?
(a) 620 kHz (b) 1530 kHz (c) 778 kHz
(d) 827 kHz (e) 1220 kHz

(a) | | (b) | | (c) | | (d) | | (e) | |

39. Monitoring a transmitter reveals the following values: plate supply voltage = 1200 V; final stage plate current = 500 mA; efficiency factor F = 70%. Antenna current = 1.9 A. Antenna resistance = 110 ohms. Using the indirect method, calculate the operating power.
(a) 600 W (b) 420 W (c) 407 W
(d) 384 W (e) 417 W

(a) | | (b) | | (c) | | (d) | | (e) | |

40. Alarm circuits and automatic control devices that are associated with broadcast station tower lights must be checked for proper operation at intervals not to exceed:
(a) one week. (b) twice each month.
(c) one month. (d) three months.
(e) six months.

(a) | | (b) | | (c) | | (d) | | (e) | |

41. What is the bandwidth of an FM broadcast channel?
(a) 10 kHz (b) 75 kHz (c) 100 kHz
(d) 150 kHz (e) 200 kHz

(a) | | (b) | | (c) | | (d) | | (e) | |

42. The RF power output from the final stage of a transmitter is 15 kW. The transmission line loss is 400 W and the antenna power gain is 3.0. What is the effective radiated power?
(a) 14.6 kW (b) 29.2 kW (c) 43.8 kW
(d) 45 kW (e) 1.2 kW

(a) | | (b) | | (c) | | (d) | | (e) | |

43. At which of the following locations must the broadcast station license be posted?
(a) At the place the licensee considers to be the principal control point of the transmitter.
(b) At the location of the announcer.
(c) In the station manager's office.
(d) At the antenna housing.
(e) None of the above is true.

(a) | | (b) | | (c) | | (d) | | (e) | |

44. An AM high level modulator uses a class "A" triode output tube with an r_p of 10 kohms. The final class "C" RF stage has a plate voltage of 600 V and a plate current of 150 mA. Calculate the approximate turns ratio of the modulation transformer.
(a) 2.24 to 1 (b) 6.25 to 1 (c) 1.58 to 1
(d) 2.5 to 1 (e) 5 to 1

(a) | | (b) | | (c) | | (d) | | (e) | |

45. The present FM broadcast standards define the band of audio frequencies that frequency modulates the main carrier as the:
(a) main channel.
(b) left stereophonic channel.
(c) right stereophonic channel.
(d) stereophonic channel.
(e) stereophonic subchannel.

(a) | | (b) | | (c) | | (d) | | (e) | |

46. A transmitter has been tuned up to (but not including) the final plate tank circuit. The next step is to:
(a) couple the antenna to the plate tank circuit and then tune the plate tank circuit to resonance. The power output will then be correct.

(b) connect the final amplifier to a dummy antenna and adjust the output coupling for the proper power output while keeping the final tank circuit in resonance. The actual antenna may then be substituted for the dummy antenna.
(c) couple the antenna to the plate tank circuit. Adjust both the plate tank circuit and the coupling adjustment for maximum antenna current.
(d) use a field strength meter and adjust both the plate tank circuit and the coupling for a maximum output.
(e) None of the above is correct.

(a) | | (b) | | (c) | | (d) | | (e) | |

47. What is the accuracy required of an antenna current meter at full scale reading?
(a) 0.5% (b) 1.0% (c) 2.0% (d) 2.5%
(e) 3.0%

(a) | | (b) | | (c) | | (d) | | (e) | |

48. How many frames per second does a television broadcast station transmit?
(a) 30.0 (b) 60.0 (c) 63.5 (d) 262.5
(e) 15, 750

(a) | | (b) | | (c) | | (d) | | (e) | |

49. In Fig. 8, what is the total impedance?
(a) 1.1 kohm (b) 5 kohms (c) 2.4 kohms
(d) 15.6 kohms (e) 2.7 kohms

(a) | | (b) | | (c) | | (d) | | (e) | |

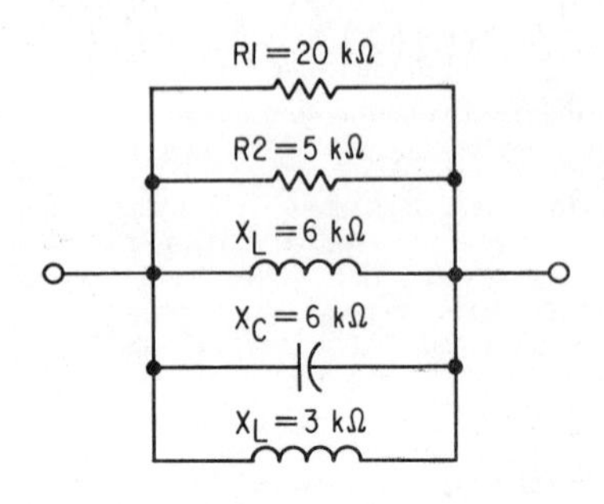

FIGURE 8

50. In Fig. 8, which component dissipates the greatest power?
(a) R1 (b) R2 (c) L1 (d) C1 (e) L2

(a) | | (b) | | (c) | | (d) | | (e) | |

Element 4, Test 8

1. The shape of the XY flexure curve of frequency change with temperature change over a -60° to +60°C range is:
(a) circular. (b) square.
(c) cubic. (d) parabolic.
(e) None of the above is true.

(a) | | (b) | | (c) | | (d) | | (e) | |

2. A broadcast television station is operating on channel 8 (180 to 186 MHz). The aural transmitter center frequency is:
(a) 180.00 MHz (b) 186.00 MHz
(c) 184.00 MHz (d) 185.75 MHz
(e) 181.50 MHz

(a) | | (b) | | (c) | | (d) | | (e) | |

3. During a "frequency test program," the broadcast station's frequency monitor indicates that the carrier frequency is 2 Hz low. During the same time period, the frequency measurement report indicates that the carrier frequency was 15 Hz high. What is the error in the station's frequency monitor?
(a) -2 cycles (b) +2 cycles (c) -17 cycles
(d) +17 cycles (e) +15 cycles

(a) | | (b) | | (c) | | (d) | | (e) | |

4. An important consideration in designing an overtone oscillator is:
(a) for the tuning to be sufficiently narrow to exclude the fundamental and other overtone frequencies of the crystal.
(b) small component size, especially above 32 MHz.
(c) that the crystal have very high resistance.
(d) that only crystals with a single electrode be used.
(e) that overtone resistors be used in the feedback loop.

(a) | | (b) | | (c) | | (d) | | (e) | |

5. An FM broadcast transmitter is 50 percent modulated by an audio frequency. What determines the rate of frequency swing of the carrier?
(a) The number of multipliers used in generating the carrier frequency.
(b) The amplitude of the audio modulating signal.
(c) The amount of deviation of the audio signal.
(d) The amount of phase shift of the audio signal.
(e) The frequency of the audio signal.

(a) | | (b) | | (c) | | (d) | | (e) | |

6. Commercial broadcast stations operated by remote control are subject to which of the following conditions?
(a) The equipment at the operating and transmitting positions shall be so installed and protected that it will not be accessible to, or capable of operation by, persons other than those duly authorized by the licensee.
(b) The control circuits from the operating positions to the transmitter shall provide positive on and off control and shall be such that the transmitter and any fault in or loss of such control will automatically place the transmitter in an inoperative position.
(c) A malfunction of any part of the remote control equipment and associated line circuits resulting in improper control or inaccurate meter readings shall be cause for the immediate cessation of operation by remote control.
(d) Control and monitoring equipment shall be installed so as to allow the licensed operator at the remote control point to perform all the functions in a manner required by the Commission's rules.
(e) All the above must be complied with.

(a) | | (b) | | (c) | | (d) | | (e) | |

7. The standard broadcast band includes the frequencies of:
(a) 25.0 to 250.0 kHz
(b) 250.0 to 535.0 kHz
(c) 535.0 to 1605 kHz
(d) 1605 to 2800 kHz
(e) 1500 to 3500 kHz

(a) | | (b) | | (c) | | (d) | | (e) | |

8. Which of the following methods may be used to adjust the "T" network to match the RF transmission line to the standard broadcast station antenna?
(a) The RF bridge method.

(b) The loading coil method.
(c) The product over the sum method.
(d) The substitution method.
(e) Both (a) and (d) are true.

(a) | | (b) | | (c) | | (d) | | (e) | |

9. A high level class "B" modulator does not:
(a) require a well regulated power supply.
(b) require greater drive than a class "A" modulator.
(c) have greater plate efficiency than a class "A" modulator.
(d) require two tubes in push-pull for minimum distortion.
(e) require less audio drive than a comparable class "A" modulator.

(a) | | (b) | | (c) | | (d) | | (e) | |

10. What type of microphone should be used in areas of high noise?
(a) The dynamic microphone.
(b) The crystal microphone.
(c) A noise cancelling microphone.
(d) The condenser microphone.
(e) The ribbon microphone.

(a) | | (b) | | (c) | | (d) | | (e) | |

11. One method of obtaining negative-current feedback in a common emitter single-ended amplifier that uses an emitter resistor for temperature stabilization is to:
(a) employ transformer coupling between the two stages.
(b) employ R-C coupling between the two stages.
(c) place a resistor in series with a dc blocking capacitor connected between the collector and the base of the transistor.
(d) leave the emitter resistor unbypassed.
(e) employ dc isolation between the two stages.

(a) | | (b) | | (c) | | (d) | | (e) | |

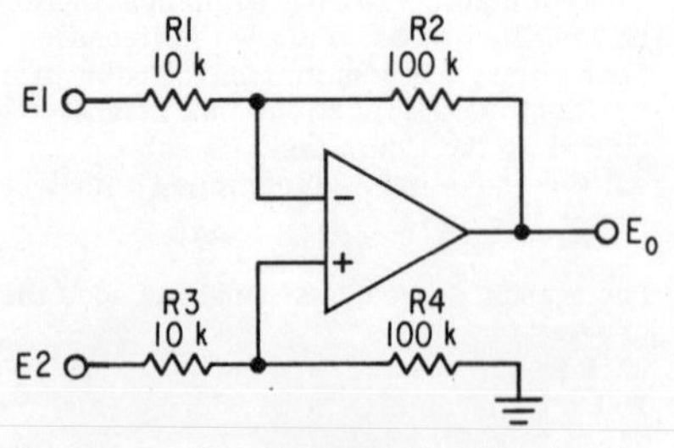

FIGURE 1

12. Refer to the operational amplifier circuit of Fig. 1. Which of the following statements is false?
(a) The input impedance to signals E1 and E2 is 10 kohms.
(b) The circuit is a differential amplifier.
(c) The gain of the circuit is given by the formula, $E_0 = -10(E1 - E2)$.
(d) If E1 and E2 are identical, any output signal would be due to a mismatch in the resistors or an unbalance in the operational amplifier.
(e) The circuit could be used as an audio mixer of signals E1 and E2.

(a) | | (b) | | (c) | | (d) | | (e) | |

13. The proximity effect in microphones is noticed by:
(a) an increase in low-frequency response.
(b) an increase in high-frequency response.
(c) a decrease in low-frequency response.
(d) a decrease in high-frequency response.
(e) None of the above are true.

(a) | | (b) | | (c) | | (d) | | (e) | |

14. Dirt or oxide build-up on the heads of a tape recorder may cause:
(a) reduction of the audio-output, especially in the high-frequency range.
(b) increased head wear.
(c) wow or flutter.
(d) increase in the noise level.
(e) All of the above are true.

(a) | | (b) | | (c) | | (d) | | (e) | |

15. The phasitron tube is used in a:
(a) modulator stage.
(b) frequency multiplier stage.
(c) buffer stage.
(d) pre-emphasis stage.
(e) final RF power amplifier stage.

(a) | | (b) | | (c) | | (d) | | (e) | |

16. If relay contacts have become severely oxidized or tarnished, but not pitted, what is the best method of cleaning them?
(a) Use a burnishing tool.
(b) Use an automotive point file.
(c) Remove the oxide with emery paper or emery cloth.
(d) Use a light grade of sandpaper.
(e) The oxidization or tarnish may be etched off with muriatic acid and the points cleaned with carbon tetrachloride.

(a) | | (b) | | (c) | | (d) | | (e) | |

17. Why are interlocks placed on access doors and grills, etc.?
(a) So that unauthorized personnel will not have access to the equipment.
(b) To protect the operating personnel from coming in contact with current-carrying conductors and components.
(c) Because interlocks are required by the FCC on certain equipment.
(d) Interlocks are no longer required on transmitting equipment.
(e) To guard against pilferage of the equipment.

(a) | | (b) | | (c) | | (d) | | (e) | |

18. The circuit shown in Fig. 2 represents:
(a) a line equalizer.

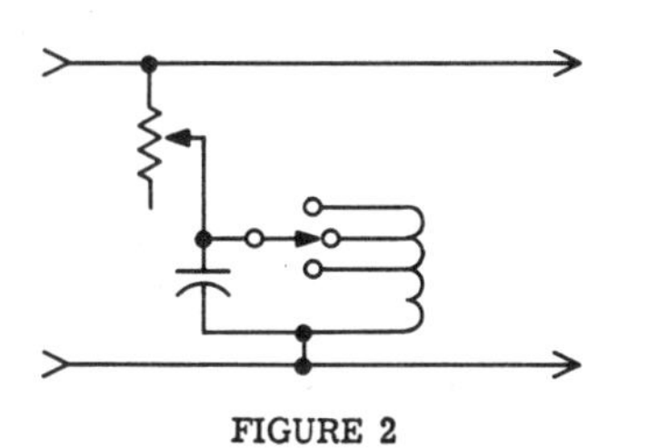

FIGURE 2

(b) a wave trap.
(c) a low pass filter for harmonic attenuation.
(d) an impedance matching network.
(e) None of the above is true.

(a) | | (b) | | (c) | | (d) | | (e) | |

19. For Radio Broadcast Services, the maximum temperature variation of the crystal from its normal operating temperature when using Y or X cut crystals is:
(a) $\pm 0.01°C$ (b) $\pm 0.10°C$ (c) $\pm 1.00°C$
(d) $\pm 2.00°C$ (e) $\pm 5.00°C$

(a) | | (b) | | (c) | | (d) | | (e) | |

20. The type of stylus generally used in broadcast phonograph pick-ups is made of:
(a) sapphire. (b) emerald. (c) diamond.
(d) steel. (e) ruby.

(a) | | (b) | | (c) | | (d) | | (e) | |

21. Which of the following is not required to be recorded and maintained in the records of a remote pick-up station?
(a) The nature and sponsorship of each program transmitted.
(b) The date and time of operation.
(c) The purpose of the operation.
(d) The station with which it communicates.
(e) Frequency check, if made.

(a) | | (b) | | (c) | | (d) | | (e) | |

22. Recycling relays may be used in a broadcast transmitter to:
(a) identify a point of overload that may occur in some of the circuits in the transmitter or on the antenna transmission line.
(b) protect the transmitter components from damage due to arc-over.
(c) keep a circuit open if an overload is continuous.
(d) extinguish an arc-over caused by lightning striking the antenna that could damage the transmission line or transmitter.
(e) All the above are true.

(a) | | (b) | | (c) | | (d) | | (e) | |

23. Neutralization of a triode RF amplifier is necessary to:
(a) increase the amplifier gain.
(b) decrease the amplifier gain.
(c) neutralize feedback from the plate circuit to the grid circuit.
(d) reduce distortion.
(e) prevent harmonic generation.

(a) | | (b) | | (c) | | (d) | | (e) | |

24. When selecting an amplifier for the speech system of an FM broadcast station, which of the following characteristics is not an important consideration?
(a) Its ability to reproduce in its output a true but amplified reproduction of the input signal over the normal audio range.
(b) Phase characteristics of the output signal as compared to the input signal.
(c) Its ability to faithfully reproduce, without distortion, a 10 Hz signal.
(d) Its freedom from hum and noise.
(e) Its ability to faithfully amplify frequencies from 0 to 200, 000 Hz.

(a) | | (b) | | (c) | | (d) | | (e) | |

25. In TV broadcast stations, the aural carrier frequency is located below the channel's upper limit by:
(a) 100 kHz (b) 150 kHz (c) 250 kHz
(d) 500 kHz (e) 1 MHz

(a) | | (b) | | (c) | | (d) | | (e) | |

26. In Fig. 3, which of the following statements would be false if R2 were open?
(a) The collector current of Q1 would decrease.
(b) The emitter-collector potential of Q_1 would be approximately 6 V dc.
(c) The collector to ground voltage of Q1 would be +12 V dc.
(d) The voltage drop across R4 would be approximately 1 V dc.
(e) The forward bias of Q1 would increase.

(a) | | (b) | | (c) | | (d) | | (e) | |

27. If a Marconi antenna is to be installed in a position remote from ground, it must be:
(a) longer than a quarter wavelength.
(b) shorter than a quarter wavelength.
(c) provided with an apparent ground or counterpoise.
(d) A Marconi antenna cannot be used remote from ground.
(e) operated at frequencies far below the frequency for which the antenna is a quarter wavelength long.

(a) | | (b) | | (c) | | (d) | | (e) | |

28. The fidelity of an audio amplifier may sometimes be improved by including:
(a) negative current feedback.
(b) H pads for controlling the amplitude of the signal.
(c) negative voltage feedback.
(d) T pads for controlling the amplitude of the signal.
(e) Both (a) and (c) are true.

(a) | | (b) | | (c) | | (d) | | (e) | |

29. In Fig. 3, if C1 is shorted, which of the following statements would be false?

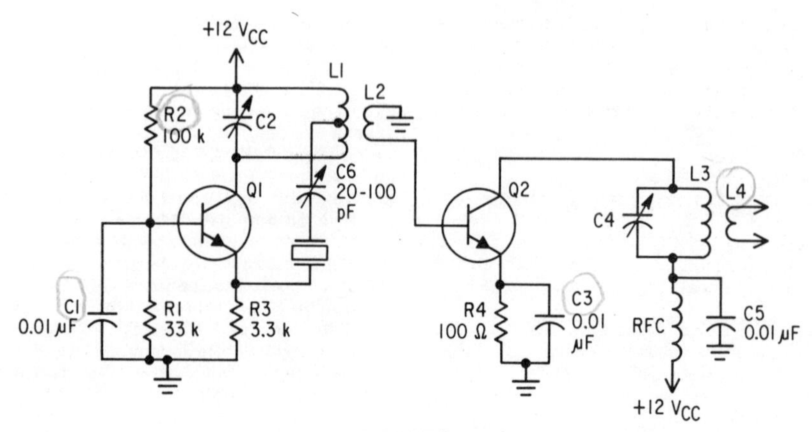

FIGURE 3

(a) The collector current of Q1 would decrease.
(b) The emitter-collector potential of Q1 would be approximately 6 V dc.
(c) The voltage drop across R3 would decrease.
(d) The voltage drop across R4 would decrease.
(e) Oscillations would cease.

(a) | | (b) | | (c) | | (d) | | (e) | |

30. A transmission line delivers 2.5 kW to an antenna whose current is measured at 3.74 A. What is the antenna's resistance?
(a) 179 ohms (b) 356 ohms (c) 676 ohms
(d) 338 ohms (e) 334 ohms

(a) | | (b) | | (c) | | (d) | | (e) | |

31. The frequency tolerance of a broadcast FM SCA subcarrier without modulation must be maintained within:
(a) ±20 Hz (b) ±100 Hz (c) ±500 Hz
(d) ±1 kHz (e) ±2 kHz

(a) | | (b) | | (c) | | (d) | | (e) | |

32. In Fig. 3, the output signal amplitude at L4 has suddenly dropped to 75 percent of its original value. This could be due to:
(a) C5 being an open circuit.
(b) C3 becoming an open circuit.
(c) the RFC becoming an open circuit.
(d) L2 becoming an open circuit.
(e) R1 becoming an open circuit.

(a) | | (b) | | (c) | | (d) | | (e) | |

33. Neutralization is not required when a triode RF amplifier is used:
(a) at ultrahigh frequencies.
(b) for low power amplification.
(c) as the modulated stage.
(d) in a push-pull circuit.
(e) as a doubler stage.

(a) | | (b) | | (c) | | (d) | | (e) | |

34. 100 percent modulation for a commercial FM broadcast station is defined as a frequency swing of:
(a) ±25 kHz (b) ±50 kHz (c) ±75 kHz
(d) ±100 kHz (e) ±200 kHz

(a) | | (b) | | (c) | | (d) | | (e) | |

35. Why are class I stations sometimes granted the use of a directional antenna?
(a) To increase their field strength in their intended service area.
(b) To protect the service area of a class III station.
(c) To protect the service area of another class I station.
(d) To conserve on the power they use.
(e) Both (a) and (c) are true.

(a) | | (b) | | (c) | | (d) | | (e) | |

36. Which of the following is not a standard reel size?
(a) 3 in. (b) 5 1/4 in. (c) 6 in.
(d) 7 in. (e) 10 1/2 in.

(a) | | (b) | | (c) | | (d) | | (e) | |

37. A broadcast transmitter is supplying 20 kW of RF power to the input of the antenna transmission line. If there is a 200 W loss in the transmission line, and the antenna has a rated power gain of 4, what is the effective radiated power?
(a) 79.2 kW (b) 19.8 kW (c) 20.0 kW
(d) 40.0 kW (e) 80.0 kW

(a) | | (b) | | (c) | | (d) | | (e) | |

38. Low-frequency record-groove velocity variations caused by either the turntable or the record are known as:
(a) rumble. (b) flutter. (c) wow.
(d) frequency response. (e) skating.

(a) | | (b) | | (c) | | (d) | | (e) | |

39. A class "C" RF power amplifier operating at 85 percent efficiency is delivering 750 W of RF power to the antenna. The RF

amplifier is being plate modulated by a class "A" amplifier operating at 25 percent efficiency and producing a 90 percent modulation of the RF carrier with a sinusoidal audio wave. What is the dc power input to the modulator?
(a) 1429 W (b) 2430 W (c) 3000 W
(d) 608 W (e) 1216 W

(a) | | (b) | | (c) | | (d) | | (e) | |

40. According to FCC standards, what are the minimum required indicating instruments for a nondirectional FM broadcast transmitter?
(a) Meters must be installed to indicate the final RF plate voltage and the plate current of all the RF stages. An RF current or volt meter must be installed to indicate the transmission line current or voltage.
(b) Plate current and plate voltage meters are required for all RF stages.
(c) Meters must be installed to indicate the final RF amplifier's plate voltage and plate current plus an RF meter to indicate the transmission line voltage or current or power.
(d) Meters must be installed to indicate the ac line voltage, ac filament voltage, dc plate voltage, and plate current of the final RF stage plus a meter to indicate the transmission line voltage or current.
(e) All stages and transmission lines must be metered.

(a) | | (b) | | (c) | | (d) | | (e) | |

41. A power amplifier has equal input and output impedances and has a power gain of 50. Express this gain in dB.
(a) 39.97 dB (b) 50.00 dB (c) 1.70 dB
(d) 3.00 dB (e) 16.99 dB

(a) | | (b) | | (c) | | (d) | | (e) | |

42. Which of the following is required to produce a full color picture with fine detail?
(a) "I" and "Y" signals.
(b) "Y" and "Q" signals.
(c) "I" and "Q" signals.
(d) "Y" and the color burst signal.
(e) Both (c) and (d) are true.

(a) | | (b) | | (c) | | (d) | | (e) | |

43. When a remote-reading RF ammeter is used to indicate the RF power being fed to the antenna, it must be checked against the regular antenna current meter at least:
(a) once per day. (b) once per week.
(c) twice per week. (d) once per month.
(e) twice per month.

(a) | | (b) | | (c) | | (d) | | (e) | |

44. The antenna system used for the aural portion of television broadcast stations is usually:
(a) randomly polarized.
(b) vertically polarized.
(c) electrostatically polarized.
(d) horizontally polarized.
(e) cross polarized.

(a) | | (b) | | (c) | | (d) | | (e) | |

45. Why are directional antennas used in some broadcast station installations?
(a) To increase the field strength of a class IV station in a desired direction in order to reduce the interference caused in that area by a class II station.
(b) To increase the secondary coverage area of a class IV station in the desired direction.
(c) To reduce the field strength in one or more directions that may cause interference to a class I-A station operating on the same or an adjacent channel.
(d) To reduce the cost of operation of the station by increasing the coverage area of the station with less RF power.
(e) All of the above are true.

(a) | | (b) | | (c) | | (d) | | (e) | |

46. The amplitude of the RF power measured during synchronizing peaks on a video TV carrier is called:
(a) authorized carrier power.
(b) effective radiated power.
(c) output power. (d) peak power.
(e) peak-to-peak power.

(a) | | (b) | | (c) | | (d) | | (e) | |

47. Except for emergencies beyond the control of AM and FM broadcast stations, the antenna input power of the station shall not deviate from the authorized input power within the limits of:
(a) +10% and -5% (b) +5% and -5%
(c) +10% and -10%
(d) less than 90% nor greater than 105%
(e) +15% and -15%

(a) | | (b) | | (c) | | (d) | | (e) | |

48. What is the bandwidth of a television broadcast channel?
(a) 6.0 MHz (b) 4.9 MHz (c) 4.7 MHz
(d) 4.5 MHz (e) 4.0 MHz

(a) | | (b) | | (c) | | (d) | | (e) | |

49. In the transmission and reproduction of broadcast color television, the "Q" signal has a bandwidth of:
(a) 0 to 4.2 MHz and contains the brightness signal that supplies the fine detail for color transmissions and the black and white information for monochrome transmission.
(b) 0 to 1.5 MHz and contains the chrominance signals for the cyan to orange color range.
(c) 0 to 0.5 MHz and contains the chrominance signals for the yellow-green to purple color range.

(d) 0 to 1.5 MHz and contains the chrominance signals for the yellow-green to purple color range.
(e) 0 to 0.5 MHz and contains the chrominance signals for the cyan to orange color range.

(a) | | (b) | | (c) | | (d) | | (e) | |

50. Negative voltage feedback is used in an amplifier to:
(a) reduce waveform distortion.
(b) increase the current gain.
(c) improve the frequency response.
(d) increase the voltage gain.
(e) Both (a) and (c) are true.

(a) | | (b) | | (c) | | (d) | | (e) | |

Element 4, Test 9

1. Under what conditions may a standard broadcast station use its facilities for communications directly with individuals or other stations?
(a) During a period of emergency or imminent emergency in the area in which the station is located, the station may use its facilities to communicate directly with individuals or other stations.
(b) If used to communicate directly with individuals or other stations, the licensee or permittee must send notice to the commission at Washington, D.C., of the action taken and the nature of the emergency as soon as possible after the beginning of such emergency.
(c) Broadcast stations are licensed to broadcast programs intended to be received by the public and must not be used to communicate directly with individuals.
(d) If the station has used its facilities to communicate directly with individuals or other stations, the licensee or permittee must send notice of the action taken and the nature of the emergency, as soon as possible after the beginning of such emergency, to the FCC Engineer in charge of the district in which the station is located.
(e) (a), (b), and (d) above are true.

(a) | | (b) | | (c) | | (d) | | (e) | |

2. Two vertical antennas are fed equal currents and are spaced 90° apart. One antenna is toward the north and the other toward the south. The current in the south antenna is lagging in phase with the current in the north antenna by 270°. What will be the horizontal radiation pattern produced?
(a) Cardioid, with a lobe toward the north.
(b) Cardioid, with a lobe toward the south.
(c) Cloverleaf, with lobes north, east, south, and west.
(d) Bidirectional, north and south.
(e) Bidirectional, east and west.

(a) | | (b) | | (c) | | (d) | | (e) | |

3. Frequency modulation systems are not used in the standard broadcast band because:
(a) of the excessive harmonic distortion generated by the FM signal.
(b) of the high power required at the standard broadcast frequencies.
(c) of the bandwidth required for the FM signal.
(d) of the audio frequency distortion produced using FM at the lower frequencies.
(e) Both (c) and (d) are true.

(a) | | (b) | | (c) | | (d) | | (e) | |

4. The RF power input to an antenna's feed point is 12,000 W, but its effective radiated power is 50 kW. The antenna's field gain is:
(a) 4.20 (b) 17.64 (c) 2.04 (d) 3 dB
(e) 0.5

(a) | | (b) | | (c) | | (d) | | (e) | |

5. Which of the following types of emission is used in the visual transmitter of a TV broadcast station?
(a) A3 (b) A5C (c) A3B (d) F3
(e) F5

(a) | | (b) | | (c) | | (d) | | (e) | |

6. An AM transmitter is being 100 percent modulated by a single frequency sinusoidal tone. The transmitter exhibits no carrier shift. What percentage of the total output power is in the sidebands?
(a) 75% (b) 50% (c) 40% (d) 33.3%
(e) 25%

(a) | | (b) | | (c) | | (d) | | (e) | |

7. What are horizontal blanking pulses used for in the US television broadcast system?
(a) To turn off the beam while it is returning to the left side of the screen.
(b) To stop the vertical oscillator while the horizontal oscillator is in operation.
(c) To control the frequency of the horizontal oscillator during the vertical blanking period.
(d) To turn off the beam during the interlace period.
(e) To start the horizontal oscillator during the retrace periods.

(a) | | (b) | | (c) | | (d) | | (e) | |

8. Recycling relays are used in broadcast stations. These may be tripped if:
(a) the current through the relay is lower

than the relay rating.
(b) a temporary overload occurs.
(c) battery charging current is excessive.
(d) harmonic radiation exceeds the allowable amount.
(e) None of the above is true.

(a) | | (b) | | (c) | | (d) | | (e) | |

9. If a certain power supply has a no load output voltage of 230 V and the regulation is 22 percent, what is the output voltage at full load?
(a) 230 V (b) 203 V (c) 189 V (d) 181 V
(e) 173.6 V

(a) | | (b) | | (c) | | (d) | | (e) | |

10. When used in a broadcast station's audio system, what purpose do limiting amplifiers serve?
(a) They compress the audio signal range.
(b) They increase the coverage of the broadcast station service area.
(c) They prevent overmodulation, which would cause distortion and adjacent-channel interference.
(d) They eliminate the need to "ride the gain" at the studio console.
(e) Both (a) and (d) are true.

(a) | | (b) | | (c) | | (d) | | (e) | |

11. The antenna system used with standard broadcast stations is usually:
(a) randomly polarized.
(b) vertically polarized.
(c) electrostatically polarized.
(d) horizontally polarized.
(e) cross polarized.

(a) | | (b) | | (c) | | (d) | | (e) | |

12. The power factor of an RF coil is found to be 0.02. Its "Q" is approximately:
(a) 20 (b) 50 (c) 200 (d) 100
(e) Its "Q" cannot be determined from the information given.

(a) | | (b) | | (c) | | (d) | | (e) | |

13. What is the audio frequency range capability of an FM broadcast station?
(a) 15 Hz to 20 kHz (b) 20 Hz to 15 kHz
(c) 20 Hz to 20 kHz (d) 50 Hz to 20 kHz
(e) 50 Hz to 15 kHz

(a) | | (b) | | (c) | | (d) | | (e) | |

14. The AT cut mode of vibration is:
(a) flexure. (b) thickness shear.
(c) length thickness. (d) face shear.
(e) length width.

(a) | | (b) | | (c) | | (d) | | (e) | |

15. When a transmission line is terminated by a nonreactive resistance equal to the line's characteristic impedance:
(a) all the power traveling down the line will be dissipated by the terminating resistance.
(b) no sharp "dips" will occur in the antenna current.
(c) the standing wave ratio on the transmission line will be very low.
(d) there will be no carrier shift.
(e) Both (a) and (c) are true.

(a) | | (b) | | (c) | | (d) | | (e) | |

16. Referring to the operational amplifier circuit of Fig. 1, which of the following statements is false?
(a) The low frequency gain E_O/E_I is -10.
(b) The input impedance is 10 k ohms.
(c) The capacitor C1 provides control of the high frequency cutoff value.
(d) The high frequency gain E_O/E_I is greater than -10.
(e) None of the above is false.

(a) | | (b) | | (c) | | (d) | | (e) | |

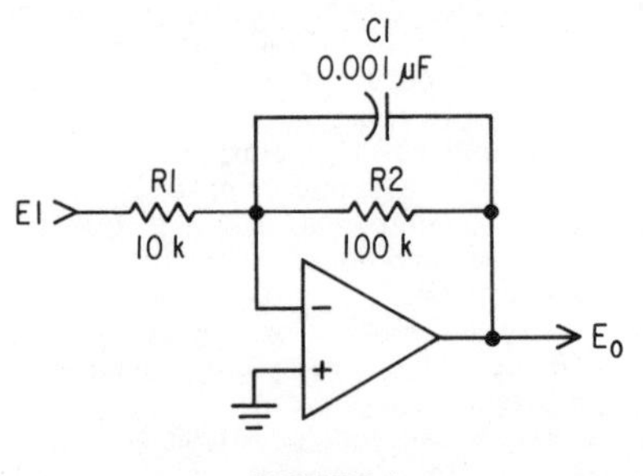

FIGURE 1

17. The characteristics of an RF class "C" power amplifier include which of the following
(a) Nonlinear operation.
(b) Low dc-to-ac efficiency.
(c) Very high collector (or plate) impedance.
(d) High power sensitivity.
(e) Very low signal-to-noise ratio.

(a) | | (b) | | (c) | | (d) | | (e) | |

18. Why is dry air or nitrogen gas sometime pumped into coaxial radio-frequency transmission lines?
(a) Dry air or nitrogen gas keeps out moistu and improves the voltage breakdown ratin of the line.
(b) A pressurized line with dry air or gas keeps out the moisture and reduces possible signal losses in the line.
(c) Pressurized lines with dry air or gas have higher surge impedances.
(d) By running a flame along such lines, leaks can be easily detected.
(e) Both (a) and (b) are true.

(a) | | (b) | | (c) | | (d) | | (e) | |

19. Figure 2 represents two horizontal dipoles, D1 and D2, which have a vertical spacing of $\lambda/2$ and are fed with equal curren in phase. Maximum radiation occurs in directions:

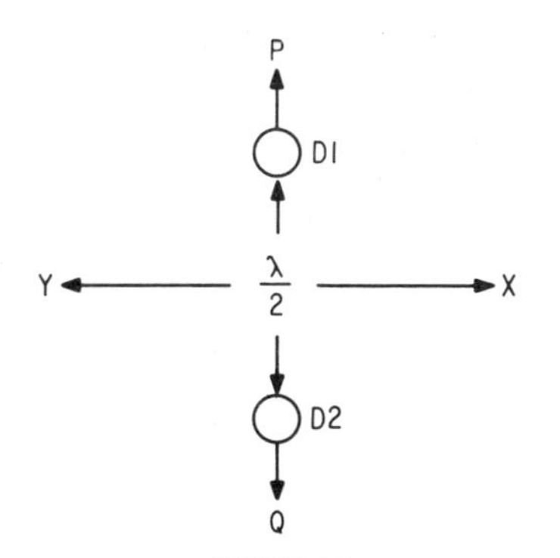

FIGURE 2

(a) P and X (b) Q and Y (c) P and Y
(d) X and Y (e) P and Q

(a) | | (b) | | (c) | | (d) | | (e) | |

20. To what percentage must the currents in the elements of a directive standard broadcast station antenna system be maintained?
(a) 10% (b) 7.5% (c) 5.0% (d) 3.5%
(e) 2.0%

(a) | | (b) | | (c) | | (d) | | (e) | |

21. In reference to Fig. 3, if a frequency change of 500 Hz is noticed when reading the station's frequency monitor, this change would not be caused by a change in the:
(a) value of C2.
(b) value of C6.
(c) value of C4.
(d) pressure of the crystal holder.
(e) inductance of L1.

(a) | | (b) | | (c) | | (d) | | (e) | |

22. Excitation energy for some broadcast station frequency monitors must be taken from an unmodulated stage of the transmitter to:
(a) keep the sideband frequencies from affecting the reading of the meter.
(b) prevent the monitor from overloading.
(c) prevent carrier shift in the transmitter.
(d) prevent the monitor from causing frequency shift of the carrier.
(e) prevent damage to the frequency monitor.

(a) | | (b) | | (c) | | (d) | | (e) | |

23. Which of the following would not be of importance when selecting a location for a VHF TV antenna?
(a) The ground plane conductivity at the antenna location.
(b) The height above the surrounding ground terrain.
(c) The line of sight to the intended coverage area.
(d) Buildings and other obstructions in the transmission path.
(e) All of the above are important.

(a) | | (b) | | (c) | | (d) | | (e) | |

24. In television transmission, there are 60 fields per second. Each field consists of a total of 262.5 lines. Of these, the number of lines actually reproducing picture information is:
(a) 242.5 (b) 262.5 (c) 252.5
(d) 232.5 (e) 212.5

(a) | | (b) | | (c) | | (d) | | (e) | |

25. A single triode audio frequency amplifier has a μ of 12. The plate voltage is 275 V, the plate current is 2.5 mA, and the ac plate impedance is 7500 ohms. If the load impedance is 30,000 ohms, what is the stage gain?
(a) 2.40 (b) 6.25 (c) 9.60 (d) 24.0
(e) 96.0

(a) | | (b) | | (c) | | (d) | | (e) | |

26. Wow and rumble would not be caused by:
(a) speed variations of the turntable.
(b) defective mechanical shock absorbers.

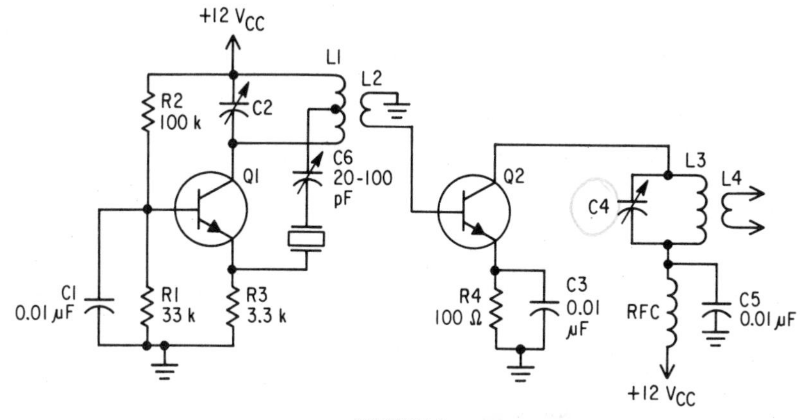

FIGURE 3

(c) a warped recording.
(d) tape skew.
(e) mechanical vibration of the turntable.

(a) | | (b) | | (c) | | (d) | | (e) | |

27. Most television transmitters employ grid modulation for the video signal. The reason for this is:
(a) grid modulation is self regulating.
(b) plate modulation cannot be used because of the horizontal sync signals.
(c) a bandwidth of over 4 MHz is used.
(d) frequency modulation is used.
(e) the plate load impedances are too high.

(a) | | (b) | | (c) | | (d) | | (e) | |

28. The load presented to the modulator stage by a plate modulated triode push-pull class "C" RF stage may be determined by:
(a) taking the square root of the product of the plate voltage and the plate current of the modulated stage.
(b) multiplying the modulator stage plate voltage by the plate current of the modulated stage.
(c) multiplying the plate voltage of the modulated stage by the plate current of that stage.
(d) dividing the plate voltage of the modulated stage by the plate current of the modulated stage.
(e) dividing the square of the plate voltage of the modulated stage by the plate current of the modulated stage.

(a) | | (b) | | (c) | | (d) | | (e) | |

29. "Secondary emission" in a vacuum tube is caused by:
(a) the emission of electrons from a tube element as a result of the impact of high velocity electrons upon its surface.
(b) electrons that are emitted due to thermal activity.
(c) the space charge cloud around the cathode.
(d) the heated cathode.
(e) electrons attracted by the control grid.

(a) | | (b) | | (c) | | (d) | | (e) | |

30. The term SCA, as used by the FCC, means:
(a) Subsidiary Communications Authority.
(b) Subsidiary Communications Authorization.
(c) sideband carrier amplitude.
(d) signal carrier amplitude.

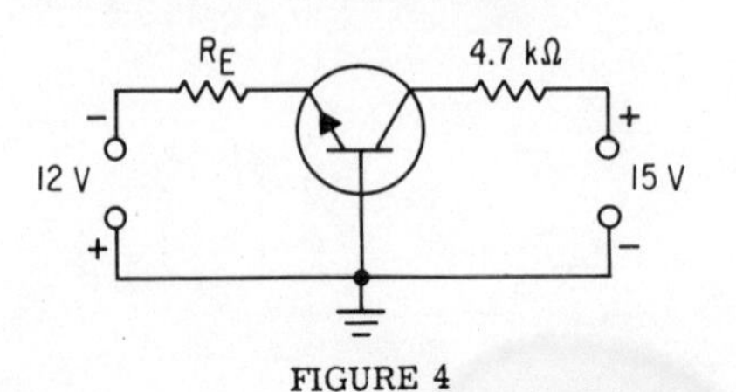

FIGURE 4

(e) suppressed carrier amplitude.

(a) | | (b) | | (c) | | (d) | | (e) | |

31. In Fig. 4, the emitter-base junction voltage drop is 0.7 V and the static transistor current gain is assumed to be unity. If the collector potential is +6.5 V, the value of R_E is approximately:
(a) 6.2 kohms (b) 6.6 kohms
(c) 4.7 kohms (d) 5.9 kohms
(e) 7.4 kohms

(a) | | (b) | | (c) | | (d) | | (e) | |

32. In a series resonant circuit, it is assumed that the total resistance in the circuit does not change. The inductance is halved but the capacitance is then adjusted to keep the resonant frequency the same. The Q of the circuit will be:
(a) Halved. (b) Doubled.
(c) Multiplied by $\sqrt{2}$. (d) Divided by $\sqrt{2}$.
(e) Squared.

(a) | | (b) | | (c) | | (d) | | (e) | |

33. The visual transmitter of a television broadcast transmitter generally uses:
(a) Cathode modulation.
(b) Phase modulation.
(c) Grid modulation.
(d) High-level plate modulation.
(e) Suppressor grid modulation.

(a) | | (b) | | (c) | | (d) | | (e) | |

34. The center frequency of the aural transmitter shall be maintained 4.5 MHz above the visual carrier frequency within the tolerance of:
(a) ±20 Hz (b) ±100 Hz (c) ±500 Hz
(d) ±1000 Hz (e) ±2000 Hz

(a) | | (b) | | (c) | | (d) | | (e) | |

35. Regarding the safety of the radio operators, which of the following would be of least importance?
(a) Overload relays.
(b) Bleeder resistors.
(c) Interlocks on the doors, grills, etc.
(d) Circuit breakers used as on-off switches.
(e) Readily available grounding hooks to discharge circuits which the operator may come in contact with during maintenance.

(a) | | (b) | | (c) | | (d) | | (e) | |

36. A reflector is used with a dipole. Relative to the direction of the main radiated lobe, the reflector is:
(a) a half wavelength long and placed a quarter wavelength behind the dipole.
(b) longer than a half wavelength and placed less than a quarter wavelength behind the dipole.
(c) shorter than a half wavelength and placed less than a quarter wavelength in front of the dipole.
(d) shorter than a half wavelength and placed

more than a quarter wavelength in front of the dipole.
(e) longer than a half wavelength and placed less than a quarter wavelength in front of the dipole.

(a) | | (b) | | (c) | | (d) | | (e) | |

37. What stage in an FM receiver reduces the signal amplitude variations?
(a) The IF filter stage.
(b) The local oscillator stage.
(c) The limiter stage.
(d) Discriminator stage.
(e) The second detector stage.

(a) | | (b) | | (c) | | (d) | | (e) | |

38. What is the frequency deviation (corresponding to 100 percent modulation) allowed in the aural transmitter of a TV broadcast station?
(a) ±5 kHz (b) ±10 kHz (c) ±25 kHz
(d) ±50 kHz (e) ±75 kHz

(a) | | (b) | | (c) | | (d) | | (e) | |

39. A thermistor has a negative temperature coefficient. This means that:
(a) the resistance of the device decreases with an increase of temperature.
(b) the resistance of the device increases with an increase in temperature.
(c) the resistance stays the same with an increase in temperature.
(d) the resistance can either increase or decrease with a change in temperature.
(e) None of the above.

(a) | | (b) | | (c) | | (d) | | (e) | |

40. An STL system is used in:
(a) satellite television transmissions to a land receiving station.
(b) program or operational communications between the studio and transmitter of broadcast stations.
(c) the VHF band of frequencies.
(d) ship-to-land communications.
(e) aircraft-to-land station communications.

(a) | | (b) | | (c) | | (d) | | (e) | |

41. In the transmission and reproduction of broadcast color television, the "I" signal has a bandwidth of:
(a) 0 to 0.5 MHz and contains the chrominance signals for the cyan to orange color range.
(b) 0 to 1.5 MHz and contains the chrominance signals for the cyan to orange color range.
(c) 0 to 4.2 MHz and contains the brightness signal for black and white (monochrome) transmission and the fine detail for color transmission.
(d) 0 to 0.5 MHz and contains the chrominance signals for the yellow-green to purple color range.
(e) 0 to 1.5 MHz and contains the chrominance signal for the yellow-green to purple color range.

(a) | | (b) | | (c) | | (d) | | (e) | |

42. A multivibrator can be used as a:
(a) harmonic generator.
(b) square wave generator.
(c) sawtooth generator.
(d) frequency divider.
(e) All of the above are true.

(a) | | (b) | | (c) | | (d) | | (e) | |

43. If a 3.58 MHz signal is mixed with a 6.52 MHz signal, which of the following frequencies will not be produced in the output of the mixer?
(a) 2.94 MHz (b) 3.58 MHz (c) 6.52 MHz
(d) 10.1 MHz (e) 10.7 MHz

(a) | | (b) | | (c) | | (d) | | (e) | |

44. The frequency tolerance of a noncommercial FM broadcast station licensed for 10 W or less is:
(a) ±1 kHz (b) ±2 kHz (c) ±3 kHz
(d) ±0.005%
(e) No frequency tolerance is enforced.

(a) | | (b) | | (c) | | (d) | | (e) | |

45. The AM standard broadcast transmitter must be designed for what percent of modulation capability?
(a) 75% to 80% (b) 10% to 85%
(c) 85% to 95% (d) 95% to 100%
(e) 100%

(a) | | (b) | | (c) | | (d) | | (e) | |

46. Turnstile antenna systems are frequently used for TV broadcast stations because:
(a) they provide gain in the horizontal plane.
(b) the directivity of the radiated horizontal plane can be produced electrically.
(c) mechanically tilting the antenna from its vertical plane can increase the field intensity below the horizontal plane.
(d) there is little loss of energy in the vertical plane.
(e) All of the above are true.

(a) | | (b) | | (c) | | (d) | | (e) | |

47. In reference to the operational amplifier circuit of Fig. 5, which of the following statements is false?
(a) The voltage gain E_O/E_I of the circuit is -1.
(b) The input impedance is very high.
(c) The output impedance is very low.
(d) The phase shift to the circuit at low frequencies is 0°.

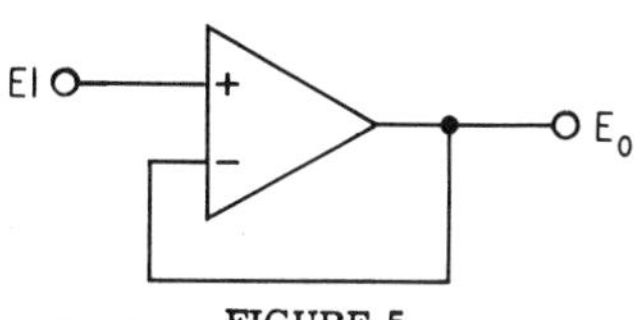

FIGURE 5

(e) The circuit is a voltage follower where $E_O = E_I$.

(a) | | (b) | | (c) | | (d) | | (e) | |

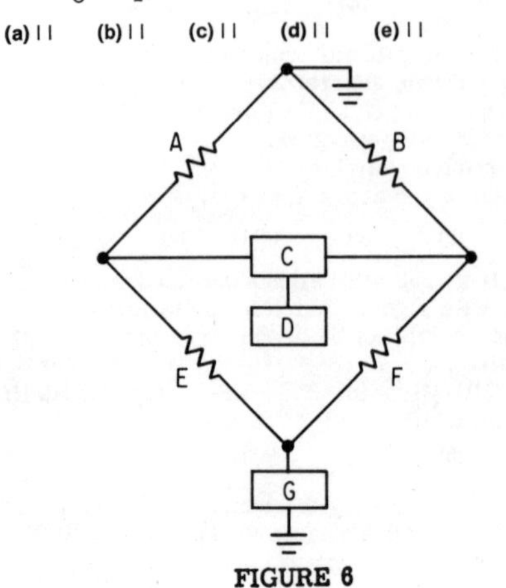

FIGURE 6

48. Figure 6 represents one method of combining the FM and video transmitter outputs to feed one antenna. The sound transmitter is normally located at point(s):
(a) C (b) D (c) G (d) either A or B
(e) either E or F

(a) | | (b) | | (c) | | (d) | | (e) | |

49. In Fig. 6, the video transmitter is normally located at point(s):
(a) C (b) D (c) G (d) either A or B
(e) either E or F

(a) | | (b) | | (c) | | (d) | | (e) | |

50. In Fig. 6, the "balun" unit would be indicated by point(s):
(a) C (b) D (c) G (d) either A or B
(e) either E or F

(a) | | (b) | | (c) | | (d) | | (e) | |

Element 4, Test 10

1. What are AGC amplifiers used for in FM, AM, and TV (aural) broadcast stations?
(a) AGC amplifiers are commonly used to stabilize the carrier frequency.
(b) They provide a flatter frequency response of the audio signal.
(c) They provide a relatively constant output amplitude for a varying input amplitude.
(d) To some extent, they eliminate the need for "riding the gain."
(e) Both (c) and (d) are true.

(a) | | (b) | | (c) | | (d) | | (e) | |

2. In reference to Fig. 1, which of the following statements is false if R3 is open?
(a) The collector current of Q1 would be zero.
(b) The voltage across R4 would be zero.
(c) The quiescent collector voltage of Q2 would be 12 V dc.
(d) The potential between the emitter-to-collector of Q2 is approximately 6 V dc.
(e) There would be no output from Q2.

(a) | | (b) | | (c) | | (d) | | (e) | |

3. Of the following statements, which is most correct concerning the power limitations imposed on STL remote pickup and intercity relay broadcast stations?
(a) The transmitted power is not to exceed 10 W ERP.
(b) The power output of the transmitter is not to exceed that necessary to render satisfactory service.
(c) The power output of the transmitter is not to exceed 5 percent of that authorized.
(d) The transmitted power is not to exceed 50 W ERP.
(e) Both (b) and (c) are true.

(a) | | (b) | | (c) | | (d) | | (e) | |

4. The signal that is fed to the erase head in a professional tape recorder is:
(a) a low frequency square wave.
(b) a low frequency triangular wave.
(c) a sine wave above audible frequency.
(d) a high frequency pulse.
(e) a 60 Hz sine wave.

(a) | | (b) | | (c) | | (d) | | (e) | |

5. Which of the oscilloscope patterns in Fig. 2 represents 75 percent modulation?

(a) | | (b) | | (c) | | (d) | | (e) | |

6. Figure 3 represents the idealized signal distribution characteristics of a TV broadcast station. The relative amplitude of the video carrier is indicated between points:
(a) D and C (b) H and G (c) A and I
(d) B and F (e) C and J

(a) | | (b) | | (c) | | (d) | | (e) | |

7. In Fig. 3, the relative amplitude of the sound carrier is indicated between points:
(a) C and J (b) C and D (c) A and I
(d) C and G (e) G and H

(a) | | (b) | | (c) | | (d) | | (e) | |

8. In Fig. 3, the characteristics of the vestigial sideband attenuation curve is indicated between points:

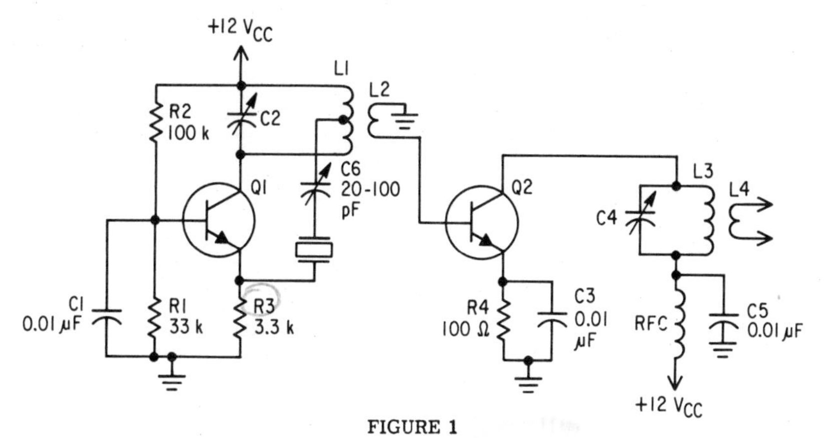

FIGURE 1

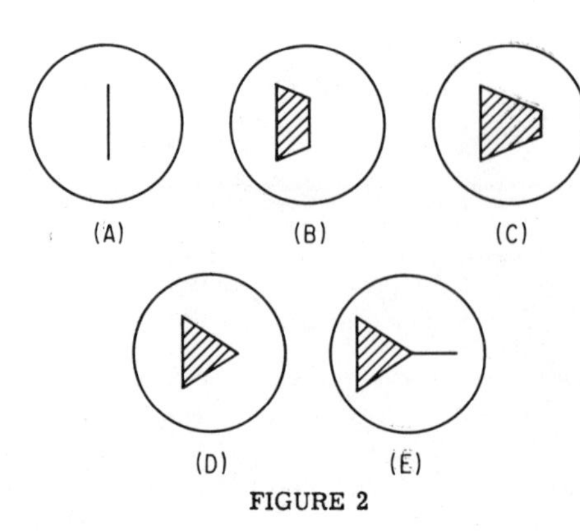

FIGURE 2

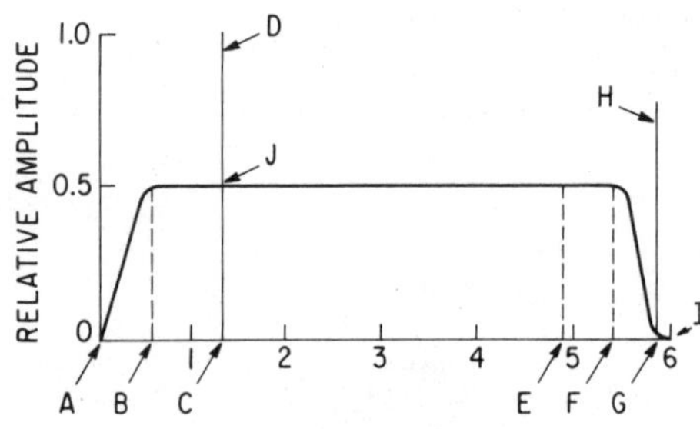

FIGURE 3

(a) B and F (b) E and G (c) A and C
(d) C and G (e) F and I

(a) | | (b) | | (c) | | (d) | | (e) | |

9. In Fig. 3, the location of the chrominance subcarrier is indicated by point:
(a) G (b) F (c) E (d) B (e) J

(a) | | (b) | | (c) | | (d) | | (e) | |

10. In Fig. 3, the bandwidth indicated between points C and E is:
(a) 4.2 MHz (b) 4.5 MHz (c) 3.58 MHz
(d) 1.25 MHz (e) 4.45 MHz

(a) | | (b) | | (c) | | (d) | | (e) | |

11. If low-temperature coefficient crystals are used in a broadcast transmitter master-oscillator circuit, the allowed temperature tolerance is:
(a) $\pm 0.05^{\circ}C$ (b) $\pm 0.1^{\circ}C$ (c) $\pm 0.2^{\circ}C$
(d) $\pm 0.5^{\circ}C$ (e) $\pm 1.0^{\circ}C$

(a) | | (b) | | (c) | | (d) | | (e) | |

12. Momentary loss of signal from a tape recording due to imperfections in the oxide on the tape is called:
(a) blocking. (b) orange peel.
(c) flutter. (d) oxide blanking.
(e) dropout.

(a) | | (b) | | (c) | | (d) | | (e) | |

13. The RMS or effective heating value of an ac sine wave is:
(a) the same as the peak value.
(b) the same as the average value taken over the complete cycle.
(c) the peak value multiplied by 0.707.
(d) the peak value divided by 1.414.
(e) Both (c) and (d) are true.

(a) | | (b) | | (c) | | (d) | | (e) | |

14. An "Austin ring" is used in broadcast station equipment:
(a) to stabilize the oscillator frequency.
(b) to block the RF current from the power line.
(c) as a safety device while performing maintenance on the transmitter.
(d) as a frequency multiplier.
(e) to provide a delay of the program for "talk" shows.

(a) | | (b) | | (c) | | (d) | | (e) | |

15. Which of the following is not true about the use of a diplexer in a broadcast TV transmission system?
(a) The diplexer can provide a 90° phase shift for nondirectional stations.
(b) The diplexer reduces vertical radiation.
(c) Phase shift networks can be included to provide directivity of the radiated horizontal field pattern.
(d) The diplexer may include a device to convert the video transmitter single-ended or unbalanced output to a balanced-to-ground output to feed transmission lines.
(e) The diplexer prevents cross-coupling between the video and sound transmitters.

(a) | | (b) | | (c) | | (d) | | (e) | |

16. The output stage of a transmitter has a plate supply voltage of 500 V, an input plate current of 500 mA, and a plate efficiency of 75 percent. If the antenna current is 3 A at the feed point and the effective antenna resistance is 22 ohms, calculate the power loss between the output stage and the antenna.
(a) 225 W (b) 198 W (c) 27 W (d) 54 W
(e) 300 W

(a) | | (b) | | (c) | | (d) | | (e) | |

17. When properly calibrated against a dummy load, the reflectometers or directional couplers may be used in TV transmission systems to:
(a) measure the power output of the transmitter.
(b) compare the incident wave to the reflected wave in order to determine the standing wave ratio.
(c) serve as a protective device for the transmitter and the transmission line.
(d) indicate problems that may exist on the transmission line or antenna.
(e) All the above are true.

(a) | | (b) | | (c) | | (d) | | (e) | |

18. During what time period may an FM station transmit signals for testing and maintenance purposes?
(a) Between 1:00 A.M. and 6:00 A.M. local standard time.
(b) Between 12:00 midnight and 6:00 A.M. local standard time.
(c) Between 1:00 A.M. and 5:00 A.M. local standard time.
(d) Between 12:00 midnight and 5:00 A.M. local standard time.
(e) None of the above is true.

(a) | | (b) | | (c) | | (d) | | (e) | |

19. As applied to FM broadcast stations, the band of frequencies from 23 to 53 kHZ containing the stereophonic subcarrier and its associated sidebands is called the:
(a) stereophonic channel.
(b) stereophonic subchannel.
(c) main channel.
(d) stereophonic subcarrier.
(e) pilot channel.

(a) | | (b) | | (c) | | (d) | | (e) | |

20. If the frequency monitor or modulation monitor of a TV, FM, or AM broadcast station becomes defective, which of the following is not required by the FCC?
(a) An entry must be made in the station's maintenance log showing the date and time the monitor was removed from service.
(b) Permission to operate without the monitor must be obtained from the FCC.
(c) The station may be operated without the monitor for a period of 60 days without further authorization from the Commission.
(d) The engineer in charge of the radio district must be notified immediately of the failure, and once again after the monitor has been restored to service.
(e) If the frequency monitor is defective, the station's carrier frequency shall be measured by an external source at least once each seven days.

(a) | | (b) | | (c) | | (d) | | (e) | |

21. How often should the antenna tower be painted?
(a) At intervals as specified in the station license.
(b) Monthly. (c) Annually.
(d) Semiannually.
(e) As often as necessary to maintain good visibility.

(a) | | (b) | | (c) | | (d) | | (e) | |

22. Regarding the measurement of the visual operating power of a TV broadcast station, which of the following statements is false?
(a) A dummy load shall be used that has virtually zero reactance.
(b) The point of measurement shall be after the vestigial sideband and harmonic filters.
(c) The output of the transmitter which includes the vestigial sideband and harmonic filters shall be terminated with a nonreactive resistance equal to the characteristic impedance of the transmission line.
(d) Electrical devices used to determine the output power shall have a full-scale accuracy of not more than $\pm$ 5 percent.
(e) During the output power measurement, the carrier shall not be modulated by any signals.

(a) | | (b) | | (c) | | (d) | | (e) | |

23. An amplifier with a gain of 65 is modified to use voltage negative feedback with a feedback factor of 20 percent. What is the amplifier's voltage gain with feedback?
(a) 13 (b) 5 (c) 6.5 (d) 4.6 (e) 7.54

(a) | | (b) | | (c) | | (d) | | (e) | |

24. The AM RF signal envelope shown in Fig. 4 represents what percentage of modulation?
(a) 66.6% (b) 72.0% (c) 75.0%
(d) 80.0% (e) 92.0%

(a) | | (b) | | (c) | | (d) | | (e) | |

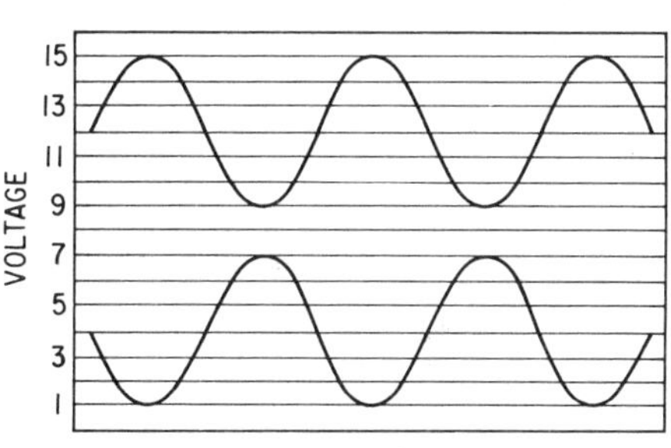

FIGURE 4

25. A standard broadcast station is operating at 1245 kHz. The antenna tower is 0.65 λ high. What is its physical height in feet?
(a) 256.0 ft (b) 488.7 ft (c) 112.7 ft
(d) 722.8 ft (e) 384.0 ft

(a) | | (b) | | (c) | | (d) | | (e) | |

26. Which of the following statements is most correct regarding the antennas used for commercial FM broadcast service?
(a) The batwing, conical, and corner reflector antennas are most commonly used.
(b) The parabolic, V dipole, and stacked ringed dipole antennas are commonly used.
(c) The stacked ringed dipole, V dipole, and curved dipole antennas are in common use.
(d) The parabolic, stacked ringed dipole, and Marconi antennas are in common use.
(e) The sheet, corner reflector, and stacked ringed antennas are commonly used.

(a) | | (b) | | (c) | | (d) | | (e) | |

27. The reactance of a 2pF capacitor is 10,600 ohms. What is the approximate frequency?

(a) 0.5 MHz (b) 0.75 MHz (c) 2.5 MHz
(d) 7.5 MHz (e) 5.0 MHz

(a) | | (b) | | (c) | | (d) | | (e) | |

28. When comparing the Image-Orthicon to the Vidicon television camera tube, which is not an advantage of the Image-Orthicon tube?
(a) Simpler tube construction.
(b) Smaller size. (c) Less weight.
(d) Lower power input.
(e) All the above are true.

(a) | | (b) | | (c) | | (d) | | (e) | |

29. A 15 kW transmitter produces a field intensity of 50 mV per meter at a position 5 miles from the transmitter. The power output of the transmitter is increased, and the field intensity is raised to 75 mV per meter. What is the new power output of the transmitter?
(a) 22.5 kW (b) 25.75 kW (c) 30 kW
(d) 33.75 kW (e) 45.25 kW

(a) | | (b) | | (c) | | (d) | | (e) | |

30. If a 72 ohm concentric line is terminated by a nonreactive resistance equal to the surge impedance of the line, what is the current flowing in the line if the power input to the line is 10 kW.
(a) 8.48 A (b) 11.8 A (c) 16.96 A
(d) 26.8 A (e) 3.73 A

(a) | | (b) | | (c) | | (d) | | (e) | |

31. In a common-base amplifier, the emitter current is 5 mA and the base current is 100 μA. The current gain is:
(a) 0.98 (b) 50 (c) 59 (d) 51 (e) 1.02

(a) | | (b) | | (c) | | (d) | | (e) | |

32. The circuit shown in Fig. 5 was designed to operate on a frequency below 5 MHz. It is an RF class "C" amplitude-modulated amplifier. What, if anything, is lacking in the circuit?
(a) A means of bias for the grid.
(b) A properly placed RF choke.
(c) A neutralization circuit.
(d) A screen-grid by-pass capacitor.
(e) Nothing is lacking in the circuit shown.

(a) | | (b) | | (c) | | (d) | | (e) | |

33. In Fig. 5, what is the value of the screen grid resistance (R_{sg})?
(a) 24.25 kohms (b) 35.00 kohms
(c) 41.25 kohms (d) 50.00 kohms
(e) 100.00 kohms

(a) | | (b) | | (c) | | (d) | | (e) | |

34. In Fig. 5, what must be the secondary impedance of the modulation transformer?
(a) 10.00 kohms (b) 8.33 kohms
(c) 7.00 kohms (d) 4.12 kohms
(e) 2.50 kohms

(a) | | (b) | | (c) | | (d) | | (e) | |

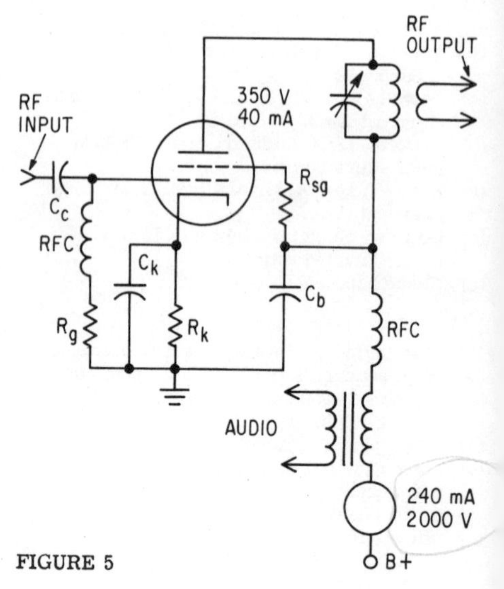

FIGURE 5

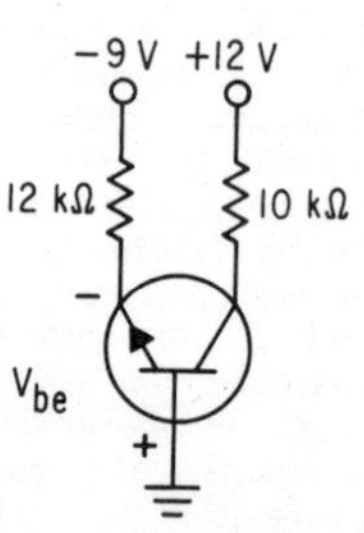

FIGURE 6

35. In Fig. 5, what is the plate efficiency of the tube if the RF output from the amplifier is 300 W?
(a) 50.0% (b) 62.5% (c) 70.0%
(d) 75.0% (e) 82.0%

(a) | | (b) | | (c) | | (d) | | (e) | |

36. In Fig. 6, the value of V_{be} may be ignored, and the static current gain is assumed to be unity. The collector potential is:
(a) +4.5 V (b) +7.5 V (c) +12 V
(d) - 9 V (e) - 7.5 V

(a) | | (b) | | (c) | | (d) | | (e) | |

37. If a light-sensitive device malfunctions that controls the tower lights, when must the lights be on?
(a) From sunset to 6:00 A.M. local standard time.
(b) From local sunset to local sunrise.
(c) From 5:00 P.M. to 6:00 A.M. local standard time.
(d) From 5:00 P.M. to sunrise local standard time.
(e) None of the above is true.

(a) | | (b) | | (c) | | (d) | | (e) | |

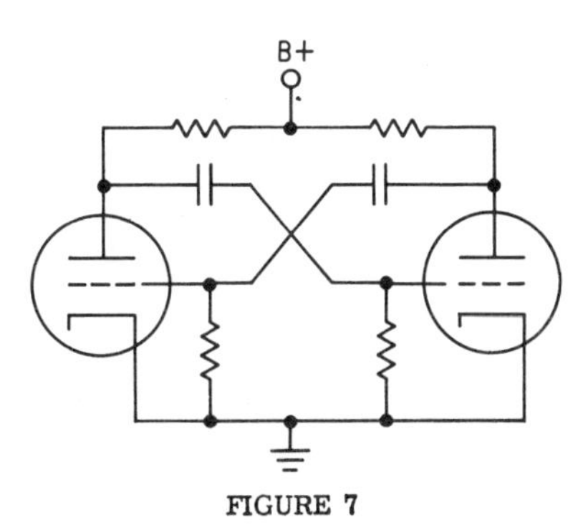

FIGURE 7

38. The circuit shown in Fig. 7 represents:
(a) a criss-cross circuit.
(b) an FM limiter.
(c) a multivibrator.
(d) a sawtooth generator.
(e) a triangular wave generator.

(a) | | (b) | | (c) | | (d) | | (e) | |

39. Two vertical antennas are fed in-phase currents and are spaced 180° apart. In which direction is the maximum radiation?
(a) The radiation pattern is perfectly circular.
(b) Along a line bisecting the towers.
(c) At right angles to the line bisecting the two towers.
(d) At 45° to the line bisecting the two towers.
(e) The pattern would be circular in the horizontal plane and hemispherical in the vertical plane.

(a) | | (b) | | (c) | | (d) | | (e) | |

40. At standard broadcast stations, tests or evidence showing that spurious radiations are suppressed must be made:
(a) daily. (b) weekly. (c) monthly.
(d) semiannually. (e) annually.

(a) | | (b) | | (c) | | (d) | | (e) | |

41. Regarding the operation and maintenance of an FM broadcast station, a second or third class radiotelephone licensee with broadcast endorsement is not allowed to:
(a) operate the switches and controls that place the transmitter on and off the air.
(b) operate the controls that compensate for voltage fluctuations in the primary power supply.
(c) make the required equipment performance measurements.
(d) operate the controls that maintain the proper modulation of the transmitter.
(e) operate the controls that turn the tower lights on and off (if the lights are manually controlled).

(a) | | (b) | | (c) | | (d) | | (e) | |

42. When used as part of the speech input equipment of a commercial broadcast station, compression amplifiers:
(a) will increase the coverage of the broadcast station's service area.
(b) will compress (or reduce) the range of the speech or music, allowing a higher percentage of modulation to be applied to the carrier.
(c) can provide an automatic announcer override of musical programs.
(d) largely eliminate the need for manually "riding the gain" at the studio console.
(e) All the above are true.

(a) | | (b) | | (c) | | (d) | | (e) | |

43. In what successive frequency steps are standard broadcast channels assigned?
(a) Every 5 kHz within the broadcast band.
(b) Every 10 kHz within the broadcast band.
(c) Every 15 kHz within the broadcast band.
(d) Every 20 kHz within the broadcast band.
(e) Every 25 kHz within the broadcast band.

(a) | | (b) | | (c) | | (d) | | (e) | |

44. Which of the following information is not required to be made available to an authorized FCC employee?
(a) The current financial statement.
(b) Program operating and maintenance logs.
(c) Equipment performance measurements.
(d) A copy of the most recent antenna resistance or common-point impedance measurements submitted to the Commission.
(e) A copy of the most recent field intensity measurements that have been made to establish performance of directional antennas.

(a) | | (b) | | (c) | | (d) | | (e) | |

45. The field strength measurement made on the radiated wave from a transmitter is 150 mV per meter at the fundamental frequency of 4.175 MHz, and 1500 μV at 8.350 MHz. What is the attenuation of the second harmonic?
(a) 20 dB (b) 40 dB (c) 60 dB
(d) 200 dB (e) 150 dB

(a) | | (b) | | (c) | | (d) | | (e) | |

46. If amplitude modulation is being used for a standard broadcast station, what is the maximum modulation percentage to be produced by the audio signal on negative peaks?
(a) Less than 50% (b) 50% (c) 75% to 95%
(d) 100% (e) 125%

(a) | | (b) | | (c) | | (d) | | (e) | |

47. A video monitor is used in TV broadcast stations to monitor two fields or less of the composite video signal. It could not be used to measure which of the following:
(a) The duty time of the vertical equalizing pulses.
(b) The peak power output of the transmitter.
(c) The time duration of the horizontal pulse interval.
(d) The amplitude of the blanking level.

(e) The amplitude of the reference black level.

(a) | | (b) | | (c) | | (d) | | (e) | |

48. Audio frequencies to be used while conducting proof-of-performance tests on an FM broadcast station include:
(a) 50, 100, 400, 1000, 5000, 10,000, and 15,000 Hz
(b) 20, 100, 400, 1000, 5000, 10,000, and 15,000 Hz
(c) 15, 50, 100, 1000, 5000, 10,000, and 15,000 Hz
(d) 50, 100, 400, 1000, 5000, and 7500 Hz
(e) 20, 50, 400, 1000, 2500, 5000, and 10,000 Hz

(a) | | (b) | | (c) | | (d) | | (e) | |

49. Which of the following is used to couple the aural and video transmitters' outputs to a common antenna?
(a) A mixer stage. (b) The diplexer.
(c) Transformer coupling.
(d) Buffer stages. (e) Duplexer stage.

(a) | | (b) | | (c) | | (d) | | (e) | |

50. When a power transformer secondary is hooked up in a Wye connection, how many single phase outputs may be obtained from its secondary?
(a) 6 (b) 5 (c) 4 (d) 3 (e) 2

(a) | | (b) | | (c) | | (d) | | (e) | |

Answers

Element 4, Test 1 Answers

1. (c) Double-humping may occur in RF transformers with both primary and secondary windings tuned. This method of coupling (Fig. 1) is commonly used in receiver IF amplifiers. The primary and secondary circuits may be identical so that both circuits have the same Q and are resonant at the IF. If the coupling factor, K(M/L), exceeds a certain critical value, equal to 1/Q, appreciable reactance will be reflected back from the secondary to the primary and will oppose and cancel some primary reactance. Then resonant "humps" will appear at frequencies F_1, F_2, which are equally distributed to either side of the IF (Fig. 2). With double-humping in the response curve, the bandwidth is equal to $\sqrt{2} \times K \times Fo$. The bandwidth produced by double-humping will be too wide for most communications receivers. However, radar receivers may use double-humping to achieve the wide bandwidth required. To reduce the bandwidth for the reception of telephony, the coupling factor must be lowered to a level near the critical value.
2. (e) (Q4.31)
3. (e) (Q4.20)
4. (c) (Q4.43) If a crystal is overdriven, it will be frequency unstable.
5. (c) (Q3.203, Q3.195, and Q4.90) RMS value of current = $\sqrt{P/R}$ = 1600/43 = 6.1 A. The peak value of current = 6.1 A x 1.414 = 8.6 A.
6. (c) (Q3.203 and Q4.57)
7. (d) (Q4.119 and R&R73.322)
8. (b) (Q3.176 and Q4.89) Total frequency multiplication factor = 2 x 3 x 3 = 18. Crystal frequency = 165.6 MHz/18 = 9.2 MHz.
9. (c) (Q4.17)
10. (e) The "Austin ring" is a power transformer that provides a low impedance path for the 60-Hz power line feeding the tower lights and provides a high impedance path to block the RF from the power line.

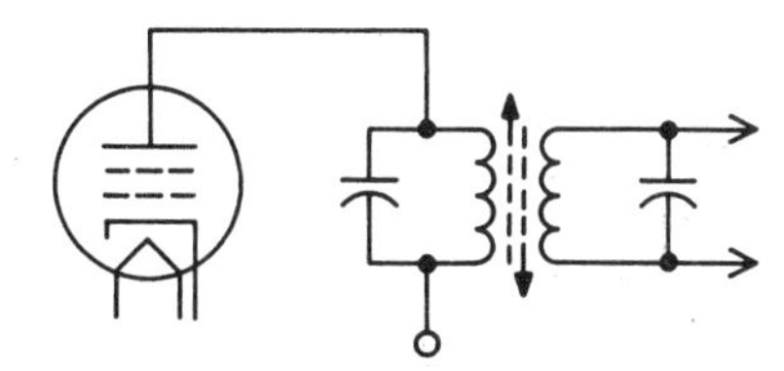

FIGURE 1

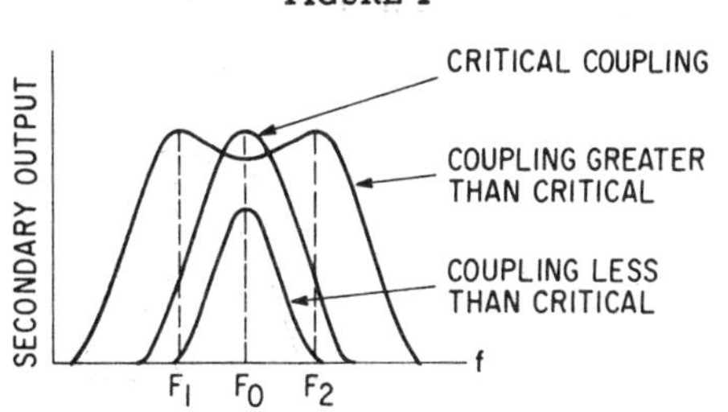

FIGURE 2

11. (e) An H pad is an attenuator and will not increase the program level.
12. (e) (Q4.17)
13. (a) (Q4.100 and R&R73.93)
14. (b) (Q4.139 and R&R17.23)
15. (d) (Q4.91 and R&R73.39)
16. (b) (Q4.73)
17. (c) (Q3.195 and Q4.90) RF power at the antenna = (Antenna current)^2 x Antenna resistance = 3.5^2 x 120 = 1470 W = 1.47 kW.
18. (d) (Q4.17)
19. (b) (Q4.124 and R&R73.689)
20. (d) (Q4.67 and Q4.68)
21. (d) (Q4.42)
22. (e) (Q3.34)
23. (b) (R&R73.902 and Q4.127)
24. (b) (Q4.36)
25. (e) (Q4.117 and R&R73.310)
26. (d) The VU meter reading will be 5 VU lower than the output from the line amp. Therefore, - 3 VU + 5 VU = + 2 VU at the

line amplifier output. The input to the telephone line = +2 VU - 6 VU = -4 VU.

27. (a) (Q3.59 and Q4.46)
28. (d) (Q4.64) RF power input to the transmission line = 80/100 x 1 kW = 800 W. RF power input to the antenna = 800 - 100 = 700 W. For an FM transmission, the RF power is independent of the modulation. Effective radiated power = 700 x 1.5 = 1050 W.
29. (a) (Q4.20)
30. (e) (Q4.137 and R&R74.561)
31. (e) (Q4.107, R&R73.113, and R&R73.114)
32. (e) (Q4.89 and R&R73.682)
33. (d) (Q3.215, Q3.216, and Q4.88)
34. (c) (Q4.83) The RMS value = 25 x 0.707 = 17.68 V.
35. (c) (R&R73.267)
36. (d) (R&R2.201 and Q4.76)
37. (c) (Q4.42)
38. (e) (Q4.24)
39. (d) (Q4.73)
40. (b) (Q3.147 and Q4.49)
41. (d) (Q4.123 and R&R73.628)
42. (a) (Q4.75 and R&R73.699)
43. (a) (Q3.192c and Q4.11) The base current, I_b = 6 mA/50 = 0.12mA. The emitter current, $I_e = I_c + I_b$ = 6.0 + 0.12 = 6.12 mA.
44. (c) (Q4.48) The total power in the sidebands = $(M^2/2) \times P_c$ where m = the modulation factor and P_c = the carrier power. The total power in the sidebands = $(0.9^2/2) \times 20 \times 10^3$ = 8.1 kW. The power in each sideband = $(8.1 \times 10^3)/2$ = 4.05 kW.
45. (b) (Q4.13) Error in frequency monitor = - 16 + (- 10) = - 26 Hz.
46. (a) The input impedance of the circuit is very high and not a direct function of resistors R1 and R2.
47. (b) (Q4.46 and Q3.129) Resonant frequency of the plate tank circuit = 2 x 1.7 = 3.4 MHz. The inductance of the plate coil, in H, is calculated as follows: $1/4\pi^2 f^2 c$ = $1/[39.4 \times (3.4 \times 10^6)^2 \times 64 \times 10^{-12}]$ = $1/[39.4 \times 3.4^2 \times 64]$ = 0.0000343 μ H
48. (a) (Q3.193) The formula for wavelength (λ) in free space is $\lambda = 300/f_{MHz}$, where f_{MHz} is the frequency in megahertz. When the wave is traveling through a conductor, its velocity will be reduced to approximately 0.94 of the free space velocity. Therefore = (300 x 0.94)/(1.5 x 4) = 47 m.
49. (c) (Q4.47) Assume an unmodulated carrier power of 100 W. For 80 percent modulation, the total sideband power = 1/2 x $(0.8)^2$ x 100 = 32 W. For 60 percent modulation, the total sideband power = 1/2 x $(0.6)^2$ x 100 = 18 W. Reduction in sideband power = 32 - 18 = 14 W. Percentage reduction (must be referred to original 32 W level) = 14/32 x 100 = 44 percent approximately.
50. (c) (Q4.01)

Element 4, Test 2 Answers

1. (d) (Q3.95 and Q4.43) As the fundamental operating frequency of a crystal increases, the thickness decreases.
2. (c) (Q4.76 and R&R73.682)
3. (c) (Q3.192A) The emitter current = I_e = (9.0 V - 0.3)/18 kohms = 8.7 mA/18 = 0.48 mA. The collector current = I_C = 0.48 x 0.95 = 0.46 mA.
4. (c) (Q3.192(M)) The collector potential = $V_C = -V_{CC} + I_C R_C = -12 + (0.46 \times 6.8) = -8.9$ V.
5. (c) (Q4.43) Series resonant oscillators are used with AT or BT cut crystals operated in the third overtone for high stability.
6. (d) (R&R73.39)
7. (c) (R&R73.96)
8. (a) The physical length of a full-wave antenna in meters = [300/f(mHz)] x 0.95 = (300/1.46) x 0.95 = 195.2 meters. For an antenna that is 0.32 wavelength long, the physical length would be 195.2 x 0.32, or 62.5 meters.
9. (a) (Q4.03) The voltage across C = 6 kohm x 12 mA = 72 V. The current through R_1 = 72 V/4 kohms = 18 mA. Current through X_L = 72 V/3 kohms = 24 mA. The total current = I_T = $\sqrt{(I_{R1})^2 + (I_L - I_C)^2} = \sqrt{(18)^2 + (24 - 12)^2}$ = 21.63 mA. The voltage across R_2 = 21.63 mA x 6.8 kohms = 147 V.
10. (b) A balanced microphone cable would require two inner conductors and one shield around them.
11. (c) (Q4.132 and R&R73.961)
12. (d) (Q4.74)
13. (a) If potentiometer R5 is not properly adjusted, there would be some off-set voltage at the output of the amplifier.
14. (b) (Q4.07)
15. (e) (Q4.07)
16. (e) 16 and 78 rpm speeds are only in limited use today due to the narrow bandwidth (16 rpm) and short playing time (78 rpm).
17. (e) (Q4.119 and R&R73.322)
18. (a) (Q3.101 and Q4.08)
19. (b) (Q4.47) Assume an unmodulated carrier power of 100 W and an antenna resistance of 100 ohms. For 80 percent modulation, total power = $P_T = P_c \times [1 + (m^2/2)] = 100\,[1 + (0.8^2/2)] = 132$ W. Antenna current = $\sqrt{P/R} = \sqrt{132/100}$ = 1.15 A. For 60 percent modulation, total power = $100\,[1 + (0.6^2/2)] = 118$ W. The antenna current = $\sqrt{118/100}$ = 1.09 A. Percentage increase = 100 (1.15 A - 1.09 A)/1.09 A = 6/1.09 A = 5.5 percent.
20. (c) (Q4.30) High frequency changes in recordings are known as "flutter," and low frequency changes once per revolution of the table or less are known as "wow."
21. (b) (Q3.83)
22. (c) (Q4.06) The new inductive reactance = 3 x 160 ohms = 480 ohms. New capacitive reactance = 160 ohms/2 = 80 ohms. Net reactance = X = 480 - 80 = 400 ohms. The total impedance = $\sqrt{R^2 + X^2} = \sqrt{(400)^2 + (400)^2}$ = 566 ohms.
23. (e) (Q4.103 and R&R73.55)
24. (e) (Q4.36)
25. (b) (Q4.122, R&R73.671, and R&R73.672)
26. (b) (Q3.215) The reflection coefficient = $\rho = \sqrt{1/10} = 0.316$. The VSWR = $(1 + \rho)/(1 - \rho)$ = 1.316/0.684 = 1.92.
27. (c) (Q4.121 and R&R73.668)
28. (c) (Q4.117, Q3.203, and Q3.203A) New field strength at the 3 mile position = 80 x $\sqrt{2}$ mV per meter. New distance of the 50 mV per meter contour from the transmitter = (80 x $\sqrt{2}$/50) x 3 = 6.8 miles.
29. (e) (Q4.35, Q4.101, R&R73.47, and R&R73.254)
30. (a) (Q4.46)
31. (b) (Q4.06) If the inductance is doubled and the capacitance is halved, the product of L x C will remain the same and the circuit will still be resonant. At resonance, the impedance is equal to the resistance of 120 ohms.
32. (e) (Q4.53)
33. (a) (Q4.25 and Q4.26)
34. (d) (Q4.52)
35. (e) For a transmission line that is an odd number of quarter wavelengths long,

the input impedance is equal to Z_o^2/Z_L $= (70)^2/125 = 39.2$ ohms.

36. (a) (Q4.141 and R&R17.25)
37. (c) (Q3.58 and Q4.05) The voltage gain of a pentode audio amplifier = $g_m \times R_L = 3500 \times 10^{-6} \times 22 \times 10^3 = 77$.
38. (b) (Q3.27 and Q4.05) The coefficient of coupling = $k = M/\sqrt{L_1 \times L_2}$, where M = the mutual inductance, in henrys, of the two coils, and L_1 and L_2 comprise the self-inductance of the two coils, in henrys.
    $k = (5 \times 10^{-3})/\sqrt{0.3 \times 15 \times 10^{-3}}$
    $= (5 \times 10^{-3})/\sqrt{4.5 \times 10^{-3}}$
    $= (5 \times 10^{-3})/(6.71 \times 10^{-2}) = 0.0745$
39. (d) (Q4.02) The fourth harmonic of 1.45 MHz = $1.45 \times 10^{-6} \times 4 = 5.80$ MHz.
40. (b) (Q4.71)
41. (d) (Q4.58)
42. (b) The output signal is inverted from the input signal (the circuit exhibits a 180° phase shift); hence the gain is -100.
43. (c) (Q4.119 and R&R73.322)
44. (b) (R&R73.767)
45. (e) Standard broadcast stations operating on the lower end of the broadcast band sometimes employ top loading to increase the effective height of the antenna, allowing the use of a shorter, more practical overall antenna height. The proper height and size of the top loading unit will improve the current distribution in the antenna and provide a lower angle of radiation.
46. (a) (Q3.147, Q4.49, and R&R73.40)
47. (d) (Q4.72)
48. (c) (Q4.102 and R&R73.51)
49. (c) (Q4.75 and R&R73.699)
50. (d) (Q3.72) The output voltage at no load = $V_{FL} \times [1 + (R/100)]$ where R = regulation percentage = $230\ (1 + 0.22) =$ 281 V.

Element 4, Test 3 Answers

1. (d) (Q4.22) It is assumed that the operator would interrupt his reading of +1 VU as the reading that occurred on frequent peaks of the varying amplitude of the music. +1 VU + 6 VU = +7 VU.
2. (c) If the potential of E1 and E2 are identical, then the differential amplifier will have no output. Any output would be due to resistor tolerances or offset within the amplifier. The potentiometer is used to provide adjustment for these offsets.
3. (c) (Q4.25 and Q4.26)
4. (a) (R&R73.682)
5. (c) Synchronizing pulses are detected in the receiver and are fed to the vertical and horizontal oscillator stages to control their frequency and phase.
6. (c) (Q3.121 and Q4.46)
7. (b) (Q3.92)
8. (a) (Q3.120 and Q4.54)
9. (e) (Q3.137 and Q4.54)
10. (a) (Q3.121 and Q4.56)
11. (b) (Q3.292 and Q4.46)
12. (a) With R3 an open circuit, there would be no screen voltage to accelerate the electrons toward the plate, and the plate and screen currents would be reduced to a very low or zero value.
13. (d) (Q4.66) The two antennas are 180° apart and are fed currents 180° out of phase. The current leaving one tower and arriving at the other antenna will be in phase and aid the radiation in that direction and vice versa.
14. (e) (Q4.45)
15. (e) (Q4.49 and Q3.147)
16. (c) (Q3.163 and Q4.63)
17. (b) (Q3.163 and Q4.63)
18. (e) Load capacitance, operating temperature and drive level must all be specified when ordering a new or replacement crystal. Without "crystal correlation" the manufacturer cannot produce the crystal unit with the proper characteristics.
19. (a) (Q4.42)
20. (a) (Q4.118 and R&R73.319)
21. (d) (Q4.74 and R&R73.682)
22. (c) (R&R2.201)
23. (e) At high frequencies, the capacitor approaches a short circuit, and the gain is +1.
24. (e) (Q4.51)
25. (e) (Q4.37, Q4.38, and Q4.39)
26. (e) (Q4.37 and Q4.38) AM transmission may use uneven positive and negative peaks to increase the station's coverage area, but the uneven positive and negative peaks applied to the FM transmitter would cause deviation beyond the allowable deviation and must be avoided.
27. (a) A high-impedance microphone cable used with some high-impedance microphones would require one inner conductor and one shield around it. Some high-impedance microphone cables use double shielding however.
28. (c) (Q4.17)
29. (b) (R&R73.628)
30. (a) (Q3.192M) The voltage drop across the 6.8 kohm resistor is 15.0 V - 8.0 V = 7 V. The collector current = I_c = 7 V/6.8 kohm = 1.03 mA. The base current = $I_b = I_c/\beta$ = 1.03 mA/35 = 29μA. The value of R_b is 15 V/29 μA = 517 kohms.
31. (c) (Q.4.07) Since a linear horizontal sweep is required to correctly reproduce the waveform being examined, a sawtooth wave must be used.
32. (a) (Q4.127, R&R73.911, and R&R73.916)
33. (c) The circuit is generally as shown in Fig. 1. Reflections back from the line, especially at high frequencies, would be

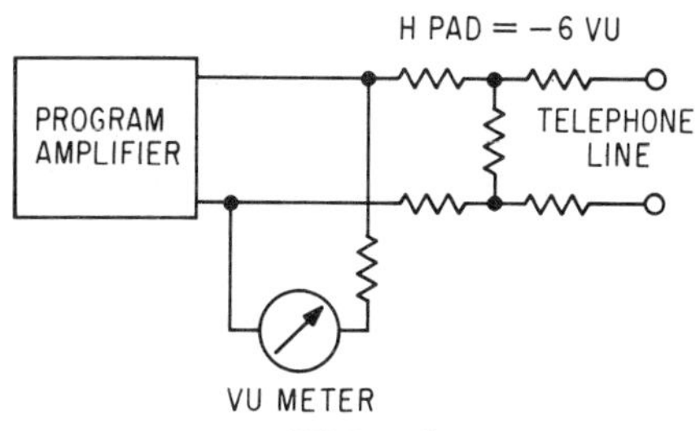

FIGURE 1

attenuated by the line pad and would prevent variations in the volume indicator readings.

34. (d) The lavalier microphone is widely used on TV shows for interviews, newscasts, or where people are moving around the set.
35. (e) (Q3.34)
36. (b) (Q4.115, R&R73.293, and R&R73.295)
37. (c) (Q3.85 and Q4.07)
38. (c) (Q4.117, Q3.203 and Q3.203A) Since field strength is directly proportional to the square root of the transmitter's power, the new field strength at the position five miles from the transmitter will be 240 x $\sqrt{3}\,\mu$V per m. Since the field strength is inversely proportional to the distance from the transmitter, the new field strength eight miles away will be 240 x $\sqrt{3}$ x 5/8 = 260 μV per m.
39. (d) (Q3.94, Q4.43, and Q4.44) The decrease in the crystal's frequency = 12 x 3 x (33°C - 30°C) = 12 x 3 x 3 = 108 Hz. Total frequency multiplication factor = 24 MHz/3 MHz = 8. Decrease in the transmitter's frequency = 108 x 8 = 864 Hz. The transmitter frequency at 33°C = 24 MHz - 864 Hz = 23.999136 MHz.
40. (c) (Q3.125, Q3.126, and Q4.54)
41. (c) (Q4.74, R&R73.681, and R&R73.699)
42. (c) (Q3.203 and Q4.50) The operating power at night = 2500 W - 1000 W = 1500 W. The operating power is proportional to $(\text{antenna current})^2$. Daytime antenna current/nighttime antenna current = $\sqrt{2500/1500}$. The daytime antenna current = 3 x $\sqrt{2500/1500}$ = 3.9 A.
43. (c) (Q4.71)
44. (a) (Q3.165)
45. (e) (Q4.14) $R_x = (R2/R1) \times R3$ = (200/100) x 300 = 600 ohms.
46. (e) (Q3.26 and Q4.05) The total inductance may be found by the formula $L_T = L_1 + L_2 + 2M$. The mutual inductance between the coils (M) = k $\sqrt{L_1L_2}$, where k is the coefficient of coupling and L_1 and L_2 are the inductances of the two coils. Therefore +2k $\sqrt{L_1L_2}$ may be used to replace 2M in the formula.
47. (c) (Q3.203) The wavelength of the radio wave whose frequency is 1320 kHz = 984 ft/1.32. The percentage = [372/(984/1.32)] x 100 = [(372 x 1.32)/984] x 100 = 49.8%.
48. (b) (Q3.06) Since $X_L = X_C$, the circuit is resonant and the total resistance = 1 kohm + 1 kohm = 2 kohms. The total conductance = 1/2 kohms = 5.0 x 10^{-4} mho.
49. (d) (Q3.203 and Q3.203A) The height of the vertical antenna in meters can be found by dividing the voltage developed in the antenna, in volts, by the field strength, in volts per meter, or 2.5/0.06 = 41.6 meters. To convert to feet, divide by 0.3048, or 41.6/0.3048 = 136.7 feet.
50. (c) (Q3.203 and Q4.57)

Element 4, Test 4 Answers

1. (b) The shotgun microphone, designed to be used with microphone booms, is highly directional. When aimed at the person speaking, it will greatly reduce the unwanted noises.
2. (e) (R&R73.112)
3. (c) (Q4.73) If both the transmitter and receiver use the same reference for their synchronizing circuits, the picture will be locked on the screen.
4. (c) (Q4.89 and R&R73.682)
5. (c) (Q4.53)
6. (b) (R&R73.682)
7. (e) (Q3.163, Q4.89, and Q4.117) The transmitter output stage is not a frequency multiplier so that its removal would not alter the output frequency (although the output power would be considerably less).
8. (e) (Q3.163, Q4.89, and Q4.117) Assuming that the design of stages 3 and 4 would allow them to be interchanged, the frequency would be unaltered, since, in both cases, their multiplying factor would be 2 x 3 = 3 x 2 = 6 times.
9. (d) (Q4.142 and R&R17.47)
10. (e) If there is no -6 V supply voltage, the transistors and diodes in the operational amplifier will not be properly biased, and the output will be a dc output approximately equal to the +6 V supply voltage.
11. (a) (Q4.05) Assume that the self-inductance of each coil is 1 H. Total inductance $= L_1 + L_2 - 2k\sqrt{L_1 \times L_2} = 1 + 1 - (2 \times 0.5 \times \sqrt{1 \times 1}) = 1 + 1 - 1 = 1$ H, or the self-inductance of one coil.
12. (b) (Q3.192C and Q4.11) The voltage drop across the 3.9 kohm resistor is 9 V - 5.5 V = 3.5 V. The emitter current $= I_E = 0.62$ V/680 ohms = 0.912 mA. The collector current $= I_C = 3.5$ V/3.9 kohms = 0.897 mA. The base current = $I_B = I_E - I_C = 0.912 - 0.897 = 0.015$ mA. The value of β is $I_C/I_B = 0.897/0.015 = 60$.
13. (e) (Q4.122 and R&R73.671 and R&R73.672)
14. (e) (Q4.117 and Q4.89) Total frequency multiplication factor = 2 x 3 x 4 = 24. In the final stage, 60 percent modulation corresponds to a frequency swing of (60/100) x 75 = $\pm$45 kHz. At the oscillator the frequency swing = $\pm$45 kHz/24 = $\pm$1.875 kHz.
15. (a) (Q4.04) An easy formula to find the required capacitance is $C_E = 10^7/2\pi f R_E$, where C_E = the capacitance, in μF; f = the lowest frequency to be passed, in Hz; and R_E = the value of the emitter resistance, in ohms. $C_E = 10^7/(6.28 \times 20 \times 1 \times 10^3) = 10^7/(1.256 \times 10^5) = 79.6\ \mu$F. The nearest practical value = 80 μF.
16. (e) In the interval of time for the current of the north antenna to travel to the south antenna, the current in the south antenna will have advanced in phase and will be in phase with the current from the north antenna; the two will be additive, increasing the radiation toward the south. The radiation from the south antenna traveling toward the north will arrive in time to be 180° out of phase and will cancel the radiation of the north antenna. As a result, a cardioid or unidirectional pattern will be produced.
17. (c) (Q3.147)
18. (e) (Q4.21 and Q4.22)
19. (c) (Q4.44)
20. (b) (Q4.30)
21. (d) (Q4.23)
22. (d) (Q4.13 and R&R73.59)
23. (a) (Q4.106 and R&R73.111)
24. (c) (Q4.28)
25. (c) (Q3.203 and Q4.50) The radiated power will increase by the square of the antenna current. Therefore, 5000 x 4 = 20,000 W.
26. (d) (R&R2.201)
27. (b) (Q3.203 and Q4.50) The transmitter power is proportional to the square of the antenna current. The new transmitter power = $2.75 \times (9.6)^2/(6.7)^2$ = 5.65 kW. Increase in the transmitter power = 5.65 - 2.75 = 2.90 kW.
28. (c) (Q4.78)
29. (e) (Q3.143, Q4.85, and R&R73.14)
30. (e) (Q4.37)

31. (a) (Q3.86) Resonant frequency = $0.159/\sqrt{L \times C} = 0.159/\sqrt{150 \times 10^{-6} \times 250 \times 10^{-12}} = 0.159/\sqrt{3.75 \times 10^{-14}} = 0.159/1.936 \times 10^{-7}$ = 821 kHz, approximately.
32. (b) (Q3.33 and Q3.34) $Q = (1/R) \times \sqrt{L/C} = (1/5) \times \sqrt{1.5 \times 10^{-4}/2.50 \times 10^{-10}} = (1000/5) \times \sqrt{0.6} = 155$.
33. (e) (Q3.33 and Q3.34) The impedance at resonance (dynamic resistance) = $L/CR = (150 \times 10^{-6})/(250 \times 10^{-12} \times 5)$ = 120 kohms.
34. (d) (Q3.33 and 3.34) The bandwidth = resonant frequency / Q (or $R/2\pi L$) = 821 kHz/155 = 5.3 kHz.
35. (e) (Q4.37 and Q4.38) Used in this position all the audio signals being transmitted will be processed by the limiting amplifier.
36. (d) This high input impedance AC amplifier has a voltage gain of +1 and extremely high input impedance.
37. (a) (Q4.81)
38. (c) (Q4.121 and R&R73.668)
39. (a) (Q4.55 and R&R73.39)
40. (b) It is highly important when ordering replacement crystals that the manufacturer be aware of the type of oscillator the crystal will be used in. For example, an AT cut crystal can be ground for a series resonant oscillator, but if placed in a parallel resonant oscillator, it would not perform well, and vice versa.
41. (b) For a section of transmission line whose length is any exact multiple of a half wavelength, the input impedance is equal to the value of the load, or 125 ohms.
42. (d) (Q3.163 and R&R73.317)
43. (e) (Q3.12 and Q4.27)
44. (c) (Q4.91 and R&R73.39)
45. (e) (Q4.74, Q4.123, and R&R73.682)
46. (c) (Q3.50, Q3.143, Q4.85, and Q4.86)
47. (c) (Q3.33 and Q3.34)
48. (e) (Q3.125, Q3.128, Q4.46, and Q4.54)
49. (d) (Q3.98 and Q4.54)
50. (c) The 100 ohms of inductive reactance resonates in series with the 100 ohms of capacitive reactance so that this combination behaves as a short circuit. The 200 ohms of inductive reactance in parallel with 200 ohms of capacitive reactance behaves as an open circuit. The source voltage is therefore equal to the voltage drop across the resistor, or 7 ohms x 2.8 A = 19.6 V.

Element 4, Test 5 Answers

1. (d) (Q4.42)
2. (b) (Q4.20)
3. (d) (Q4.74, R&R73.682, and R&R73.699)
4. (d) (Q3.88)
5. (b) (Q3.06) Conductance (G) = 1/2000 ohms = 5×10^{-4} mho. Susceptance (B) = 5×10^{-4} mho. Admittance (Y) = $\sqrt{G^2 + B^2}$ = $\sqrt{(5 \times 10^{-4})^2 + (5 \times 10^{-4})^2} = 7.07 \times 10^{-4}$ mho.
6. (b) (Q3.203 and Q4.57)
7. (a) (Q3.163, Q3.21, and Q4.63) Time constant (T) = RC, where T is in seconds, R is in ohms, and C is in farads. C = T/R = $(25 \times 10^{-6})/(1 \times 10^5)$ F = 250 pF.
8. (d) (Q4.20) The total gain in dB = 64 + 12 + 4 = 80 dB. The formula, NdB = $20 \log_{10} (E_2/E_1)$, may be used in its transposed form in solving for the gain, or $(E_2/E_1) = \text{antilog}_{10} (\text{NdB}/20)$ = $\text{antilog}_{10}(80/20)$ = 10,000.
9. (e) The circuit provides a means of linearly summing (or mixing) the two input signals.
10. (c) (Q4.90 and R&R73.14)
11. (b) (Q4.64) Antenna power gain = $10 \log_{10}$ (60 kW/15 kW) = $10 \log_{10}$ (4) = 10 x 0.6 = 6 dB.
12. (b) (Q4.17)
13. (e) (Q4.17)
14. (b) (Q3.192M and Q4.11) The voltage drop across the 150 kohm resistor = 20 - 0.3 = 19.7 V. The base current = 19.7 V/150 kohms = 0.1313 mA = 131 μ A, rounded off.
15. (b) (Q3.192C) The collector current = $\beta \times I_b$ = 40 x 131.3 μ A = 5.25 mA. The collector potential = - 20 + (5.25 x 3.3) = -2.7 V.
16. (b) (R&R73.569)
17. (e) (Q4.144 and R&R17.47)
18. (b) (Q4.25) The percentage feedback (β) = (1/a') - (1/a), where a' = the gain with feedback and a = the gain without feedback. Substituting, β = (1/15) - (1/20) = 0.067 - 0.05 = 0.017 = 1.7%.
19. (b) When microphones are used in very high sound pressures (such as picking up the music of a rock and roll band), the output of the microphone may overdrive the preamplifier input and be distorted. Pads are sometimes inserted between the microphone output and the input to the preamplifier to prevent this distortion.
20. (a) An H or T pad cannot provide dc isolation between two circuits.
21. (a) (Q4.102 and R&R73.51)
22. (e) (Q4.90 and R&R73.14)
23. (b) (Q4.75, Q4.89, and R&R73.689)
24. (d) (Q3.203 and Q4.50)
25. (d) (Q4.59)
26. (d) (Q3.213 and Q4.50) Power input to the transmission line = $E^2/Z_o = (1250)^2/500$ = 3125 W. Power input to the antenna = $I^2 \times Z_0 = (2.435)^2 \times 500$ ohms = 2965 W. Power loss = 3125 - 2965 = 160 W.
27. (d) (Q3.163) Deviation ratio = Frequency deviation/Highest audio frequency = 20 kHz/8 kHz = 2.5.
28. (d) (Q3.213) Since the length of the line is an odd multiple of a quarter wavelength, the surge impedance, $(Z_0) = \sqrt{Z_L \times Z_{in}} = \sqrt{75 \times 27}$ = 45 ohms.
29. (d) The gain of the amplifier is 11. Therefore, 0.1 V RMS x 1.414 x 11 = 1.55 V peak, which is the normal output.
30. (c) (Q4.89 and R&R73.682)
31. (e) (Q3.183)
32. (a) (Q4.74, R&R73.682, and R&R73.699)
33. (e) The schematic of the T and H pads are shown in Fig. 1. The T pad is used for unbalanced to ground systems and the H pad is used for balanced to ground systems. When the control is turned clock-

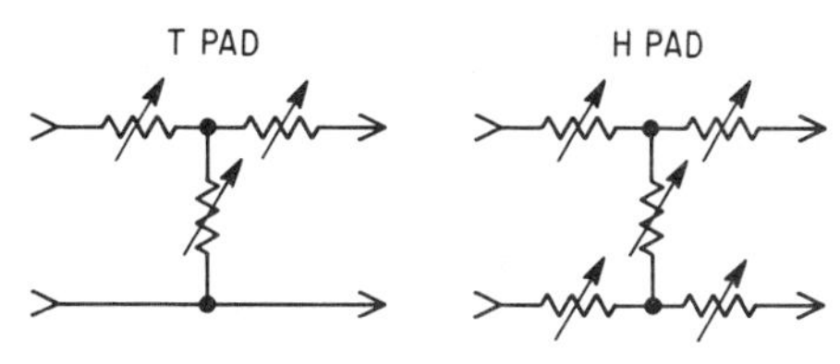

FIGURE 1

wise, the series resistance increases, while the parallel resistance decreases, causing a larger voltage division between the input and output circuits and thus greater attenuation. As the control arm is turned counterclockwise, the series resistance reduces in value as the parallel resistance increases, causing less voltage division and thus less attenuation of the output signals. If proper resistance values are chosen, the input and output impedances will not be affected.

34. (b) (Q4.19)
35. (a) (Q4.53)
36. (b) (Q4.53)
37. (c) (Q3.52, Q3.57, and Q3.59) DC power input to the amplifier = 750 V x 1.3 A = 975 W. Plate dissipation = 975 x[1 - (72/100)]= 273 W.
38. (d) If there is no +6 V supply voltage, the transistors and diodes of the operational amplifier will not be properly biased and the output will be a dc output approximately equal to the - 6 V dc supply voltage.
39. (e) (Q3.96 and Q4.54)
40. (b) (Q4.66) The radiation pattern would be similar to that of 180° spacing, driven in phase with each other, except that the null that appears between the two towers would be less of a null.
41. (d) (Q4.114 and R&R73.269)
42. (a) (Q4.123 and R&R73.681)
43. (c) (Q3.124 and Q4.46)
44. (b) (Q4.20)
45. (c) (Q4.20)
46. (b) (Q4.90 and Q4.102) Current $I = \sqrt{P/R} = \sqrt{10.08 \text{ kW}/70 \text{ ohms}} = \sqrt{144} =$ 12 A.
47. (d) (Q4.74 and R&R73.682)
48. (e) (Q3.147 and Q4.86)
49. (c) (Q3.224 and Q4.57 and Q4.117)
50. (d) (Q4.37 and 4.38) A broadcast station limiting amplifier is an amplifier that prevents overmodulation of an AM transmitter on program peaks.

Element 4, Test 6 Answers

1. (e) (Q3.192M) The base potential is (20 x 12)/(12 + 27) = 240/39 = 6.15 V. The approximate emitter current is 6.15 V/10 k = 0.62 mA.
2. (c) (Q3.192C and Q3.192G) The emitter potential = +6.15 V. Since β = 100 and $\alpha = \beta/(1 + \beta)$, α = 100/101, which is virtually 1. The collector current is 0.62 mA. The collector potential is +(20 - 6.8 x 0.62) = +15.78 V. The collector/emitter voltage is 15.78 - 6.15 = 9.6 V.
3. (e) (Q4.40 and Q4.41)
4. (e) (Q4.102 and R&R73.51)
5. (d) (Q4.91 and R&R73.39)
6. (b) (Q3.09, Q4.03, and Q4.06) Find the current in each branch. $I_{XL} = E/X_L$ = 240/40 = 6 A. $I_{XC} = E/X_C$ = 240/32 = 7.5 A. $I_R = E/R$ = 240/20 = 12 A. The total current may be found by using the law of squares, $I_T = \sqrt{(I_R)^2 + (I_{XC} - I_{XL})^2} = \sqrt{(12)^2 + (7.5 - 6)^2} = \sqrt{144 + 2.25}$ = 12.09 A.
7. (b) (Q4.42 and Q4.43)
8. (e) (Q4.44)
9. (a) (Q4.75)
10. (b) (Q4.77)
11. (a) (Q3.218 and Q4.88)
12. (d) (Q4.76 and R&R73.682)
13. (c) (Q4.82)
14. (a) (Q4.17)
15. (b) (Q4.48) Assume an unmodulated carrier power of 100 W. The total sideband power = $1/2\ m^2P_c = 1/2 \times (0.8)^2 \times 100$ = 32 W. The percentage of total sideband power in relation to the unmodulated carrier power = (32/100) x 100 = 32%.
16. (e) (Q4.17)
17. (a) (Q4.53)
18. (a) (Q4.89 and R&R73.682)
19. (c) (Q3.74)
20. (b) (Q4.127 and R&R73.915)
21. (b) Power output = $E^2/R = (55)^2/500$ = 6.05 W.
22. (d) (Q4.18) When two or more microphones are used to pick up the same program material, it is essential that they be correctly phased to prevent such effects as nondirectional pickup and reduced output and distortion.
23. (d) (Q4.17)
24. (c) (Q4.32)
25. (e) The wave radiated from the north tower would arrive at the south tower sufficiently out of phase to partially cancel its wave, and the wave traveling from the south antenna would arrive at the north antenna to partially add to the radiation in the north direction.
26. (e) (Q4.145 and R&R17.50)
27. (b) For speech and music the 1/4" standard tape width is more commonly used. The standard tape width for professional recorders is 1/2".
28. (e) (Q3.125 and Q4.85)
29. (b) The secondary impedance of the transformer must match the plate resistance of the class C modulated stage and may be found by Ohm's law: Z = E/I = 5000 V/2.5 A = 2.0 kohms, where Z = the secondary impedance, in ohms; E = the dc plate supply voltage, in volts; and I = the dc plate current, in amps.
30. (d) (Q3.276, Q3.296 and Q4.85) C5 is an RF by-pass capacitor. It provides a low-impedance path for the RF signal to return to the cathode.
31. (a) (Q3.143)
32. (c) (Q4.102) The plate efficiency of the RF modulated stage = (RF power output/dc power input) x 100 = (10 kW/12.5 kW) x 100 = 80%.
33. (d) (Q4.48) The audio power required in the secondary of the modulation transformer = dc plate power of the RF stage/2 = 12.5 kW/2 = 6.25 kW. Since the efficiency factor is 65%, the required dc input power to the moderator stage = 80%.
34. (c) To maintain constant tape speed and to provide motion to the tape, the pinch roller and capstan method is commonly used.
35. (c) (Q4.81)
36. (a) (Q3.163) In a frequency modulated transmitter, the amplitude of the carrier

is unchanged during modulation. Therefore, the transmission line current during modulation would still be 10 A.

37. (d) (Q4.117 and R&R73.310)
38. (e) (Q4.20 and Q4.25) When the open loop gain of the amplifier is extremely high, the actual gain of the circuit is approximately the ratio of R2 to R1, or 100 k/1 k = 100. Gain in VU = 20 $\log_{10}$ (V gain) = 20 $\log_{10}$ (100) = 20 x 2 = +40 VU approximately.
39. (b) (Q4.56 and R&R73.39)
40. (b) (Q3.129)
41. (a) (Q4.24) The signal fed to the program mixer would normally have a low amplitude if not first passed through a preamplifier. Since the program mixer generates some noise, this arrangement could result in a poor signal-to-noise ratio at its output. The preamplifier prevents this by increasing the signal level before it is fed into the program mixer.
42. (e) An FM receiver requires audio high frequency de-emphasis to have a resultant flat frequency response.
43. (d) The power gain of the antenna increases in proportion to the number of bays, and the power gain is approximately the number of bays in use.
44. (c) (Q4.103 and R&R73.55)
45. (e) (Q3.73 and Q4.83)
46. (e) (Q3.205 and Q4.70)
47. (a) (Q4.50) The impedance at the feed point to the antenna = $P/I^2 = (50 \times 10^3)/(10)^2$ = 500 ohms. The new current for an operating power of 5 kW = $\sqrt{P/R} = \sqrt{(5 \times 10^3)/500}$ = 3.16 A.
48. (e) (Q4.46) Grid-leak bias varies automatically with the amplitude of the input signal. If the input signal amplitude decreases, so does the bias, and this relationship tends to maintain the value of the average plate current. Similarly, any increase of input signal will cause an increase of bias, again tending to maintain the value of the average plate current.
49. (e) (Q3.94, Q4.43, Q4.44, and Q4.93)
50. (b) (Q4.63 and Q3.163)

Element 4, Test 7 Answers

1. (d) (Q4.42 and Q4.43)
2. (a) (Q4.74 and R&R73.699)
3. (b) (Q4.74 and R&R73.699)
4. (b) (Q4.74, Q4.75, and R&R73.699)
5. (d) (Q4.74, Q4.75, and R&R73.699)
6. (b) (Q4.75 and R&R73.699)
7. (e) (Q4.75 and R&R73.699)
8. (c) (Q3.93, Q3.94, Q4.43, and Q4.44)
9. (c) (Q3.192M) A class "A" amplifier operates with forward bias. For class "B" amplification, the forward bias is near zero while a class "C" stage frequently operates with zero bias.
10. (e) (Q3.183 and Q3.192K) When the slider is moved towards R7, the base of Q4 will become more positive, causing it to conduct more. The collector will be less positive and will reduce the forward bias of Q3 and also of the regulator transistor, reducing its current.
11. (c) (Q3.182 and Q3.192K) R8 is used to set the output voltage.
12. (c) (Q3.183 and Q3.192K) With no dc input at pin 2, Q1 is biased off. Q2 base is now more positive, and this transistor does not conduct.
13. (a) (Q3.183 and Q3.192K) If the voltage at point 10 becomes more positive, this will increase the conduction of Q4, decrease the conduction of the driver Q3, and decrease the conduction of the regulator transistor, thus reducing the output voltage at pin 13.
14. (d) (Q3.183 and Q3.192K) When +13.6 V is applied to input pin 2 by the transmitter "press-to-talk" switch, Q1 is biased on (base bias is derived from the voltage divider R3-R4). When Q1 conducts, the base voltage for Q2, derived from the junction of R1 and R2, becomes less positive, biasing Q2 into conduction.
15. (b) (Q4.18)
16. (d) (Q4.89)
17. (e) Capacitor C6 could be omitted and the crystal could be wired to the tap on L1, but the output frequency would change slightly.
18. (b) (Q3.178 and Q4.59)
19. (b) (Q4.64 and R&R73.310)
20. (e) (Q3.13 and Q3.40) A step-down transformer may be used to match a high primary impedance to a low secondary impedance. Impedance ratio = $(25)^2$ = 625 to 1.
21. (a) (Q4.15)
22. (a) (Q4.66) As compared to two towers with 180° spacing fed 180° apart, the 90° spacing will produce a pattern similar but with slightly narrower lobe widths.
23. (d) (Q4.06) The new value of inductive reactance = 80 x 0.707 ohm = 56.56 ohms. The new value of capacitive reactance = 80/0.707 ohm = 113.15 ohms. Net reactance = 113.15 - 56.56 = 56.59 ohms. Impedance of circuit = $\sqrt{80^2 + 56.59^2}$ = 98 ohms.
24. (d) (Q4.03 and Q4.06) Step 1: Find the total impedance (Z) of the series circuit. $Z = \sqrt{R^2 + (X_L - X_c)^2} = \sqrt{(5)^2 + (20 - 10)^2} = \sqrt{25 + 100} = \sqrt{125} = 11.18$ ohms. Step 2: Find the total voltage across the series components. $E_T = I_T \times Z_T$ = 1.5 A x 11.18 ohms = 16.8 V.
25. (d) 33 1/3 rpm is a common record player speed but not a standard speed for tape recorders.
26. (c) (Q3.246)
27. (b) (Q4.67 and Q4.68)
28. (d) (R&R73.59 and Q4.97)
29. (d) The integrator circuit produces a triangular wave output with a decreasing signal during the positive excursion of the square wave input signal.
30. (e) (Q4.33)
31. (e) (Q4.40 and Q4.41)
32. (c) (Q3.94, Q4.43, Q4.44, and Q4.93)
33. (a) (Q4.81)
34. (a) (Q3.163) The power output from an FM transmitter is constant and independent of the modulation. The total RF power output therefore remains at 50 kW.
35. (a) (Q3.203 and Q4.66) Wavelength corresponding to 925 kHz = 984/0.925 = 1063.78 ft. A 120° separation is equiva-

lent to 120/360, or one-third of a wavelength. The distance between the towers = (1/3) 1063.78 = 354.6 ft, approximately.

36. (b) (Q3.09C and Q4.03) Method 1: Total reactance of X_L and X_C in parallel. X = (12 x 15)/(15 - 12) = 60 ohms. Total impedance = $\sqrt{R^2 + X^2} = \sqrt{75^2 + 60^2}$ = 96 ohms. Method 2: Assume a voltage of 60 V across the parallel combination. Inductor current = 60 V/15 ohms = 4 A. Capacitor current = 60 V/12 ohms = 5 A. Current through resistor = 5 A - 4 A = 1 A. Voltage across the resistor = 1 A x 75 ohms = 75 V. Source voltage = $\sqrt{60^2 + 75^2}$ = 96 V. Total impedance = 96 V/1 A = 96 ohms.

37. (a) (Q3.94 and Q4.93) The minus sign before the 12 indicates that the temperature coefficient of the crystal is negative or that the operating frequency change is inversely proportional to the temperature. Therefore: 12 x (50 - 45) x 1.5 = 90 Hz. 1,500,000 - 90 = 1,499,910 Hz.

38. (c) (Q3.33 and Q3.86) Resonant frequency is inversely proportional to $\sqrt{C}$. New resonant frequency/975 kHz = $\sqrt{187/294}$. New resonant frequency = 975 x $\sqrt{187/294}$ = 778 kHz, approximately.

39. (b) (Q4.102) Using the indirect method, operating power = 1200 V x 500 mA x (70/100) = 420 W.

40. (d) (Q4.143 and R&R17.47)

41. (e) (Q4.117 and R&R73.310)

42. (c) (Q4.64) RF power arriving at the antenna feed point = 15 kW - (400/1000) kW = 14.6 kW. Effective radiated power = 3 x 14.6 = 43.8 kW.

43. (a) (Q4.99 and R&R73.92)

44. (a) (Q3.33) Load represented by class "C" stage = 600 V/150 mA = 4 kohms. For maximum undistorted output, triode load must be approximately 2 x r_p = 20 kohms. The turns ratio required = $\sqrt{20 \text{ kohms}/4 \text{ kohms}}$ = 2.24 approximately.

45. (a) (Q4.117 and R&R73.310)

46. (b) (Q3.176 and Q4.113) If a dummy antenna is available, it should be used before power is applied to the actual antenna.

47. (c) (R&R73.39 and Q4.91)

48. (a) (Q4.73 and R&R73.682)

49. (c) (Q4.03) Method 1: Reactance of X_{L1} and X_{L2} in parallel = (6 x 3)/(6 + 3) = 2 kohms. Total reactance of circuit = (2 x 6)/(6 - 2) = 3 kohms. Total resistance of circuit = (20 x 5)/(20 + 5) = 4 kohms. Total impedance = (3 x 4)/$\sqrt{3^2 + 4^2}$ = 12/5 = 2.4 kohms. Method 2: Assume 60 V to be applied across the circuit. Total reactive current = (60/6) + (60/3) - (60/6) = 20 A. Total resistive current = (60/5) + (60/20) = 15 A. Total line current drawn from 60 V source = $\sqrt{15^2 + 20^2}$ = 25 A. Total impedance = 60 V/25 A = 2.4 kohms.

50. (b) (Q3.09 and Q4.03) For all the reactive components, the mean power over the cycle is zero. In a parallel circuit, the lowest value of resistance dissipates the most power. Hence R2 dissipates four times the power dissipated by R1.

Element 4, Test 8 Answers

1. (d) The flexure curve is shown in Fig. 1.
2. (d) (Q4.76, R&R73.603, and R&R73.682) The video carrier is located 1.25 MHz above the lower channel frequency limit, and the aural center carrier frequency is located 4.5 MHz above the video transmitter carrier frequency. Thus, 180 MHz + 1.25 MHz + 4.5 MHz = 185.75 MHz.
3. (c) (Q4.13) The frequency monitor reads - 2 Hz, while the carrier frequency was 15 Hz above the assigned carrier frequency. Thus the error is 2 + 15 = 17 Hz low.
4. (a) (Q3.95)
5. (e) (Q3.162)
6. (e) (Q4.98 and R&R73.67)
7. (c) (R&R73.2 and Q4.90)
8. (e) (Q4.56, Q4.82, and R&R73.54)
9. (e) (Q3.52, Q3.111, Q4.46, and Q4.85)
10. (c) Special microphones have been designed to use in noisy areas such as sports events to reduce the background noise. Since the announcer must work close to these microphones, special wind screens and baffles have been added to reduce "popping" when his breath hits the microphone.
11. (d) (Q3.187 and Q3.192)
12. (a) The input impedance at E1 is 10 kohms and the input impedance at E2 is 110 kohms.

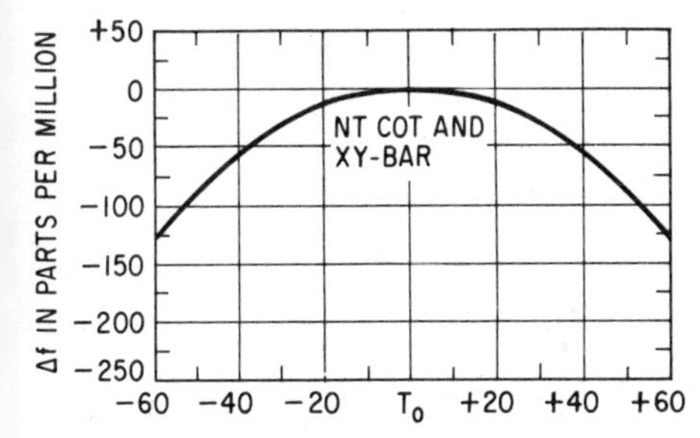

FIGURE 1

13. (a) The proximity effect in a microphone is the increase in the low frequency response when the distance from the sound source is decreased. It is most noticeable at distances of less than 2 feet.
14. (e) (Q4.29)
15. (a) (Q4.62)
16. (a) (Q4.52)
17. (b) (Q4.51)
18. (a) (Q4.36)
19. (b) (Q4.93 and R&R73.40)
20. (c) (Q4.28)
21. (a) (Q4.134 and R&R74.81)
22. (e) (4.53)
23. (c) (Q3.125, Q4.86, and Q4.85)
24. (e) (Q4.27)
25. (c) (Q4.76 and R&R73.699 Fig. 5)
26. (d) (Q3.192U) If R2 is open, Q1 will not be forward biased, and oscillations will cease. There would be no drive to Q2, and no self-bias could be developed across R4.
27. (c) (Q3.204 and Q4.70)
28. (e) (Q4.26)
29. (b) If C1 were shorted, Q1 would lose its forward bias. Oscillations would cease, and the emitter-to-collector potential would be approximately that of the supply voltage.
30. (a) (Q4.102 and R&R73.51) Antenna resistance = P/I^2 = 2500 W/(3.74 A)^2 = 179 ohms.
31. (c) (R&R73.319)
32. (b) (Q3.192U) If R4 is not by-passed, degeneration will occur and the output will be reduced.
33. (e) (Q3.131)
34. (c) (Q4.117 and R&R73.310)
35. (e) A clear channel station may be granted the use of a directional antenna in order to provide satisfactory service in their service area, or to avoid interference to another clear channel station's service area operating on the same frequency.
36. (c) The 6-in. reel size is a nonstandard size.
37. (a) (Q3.213 and Q4.64) The effective

radiated power (ERP) = (20,000 - 200) x 4 = 79.2 kW.

38. (c) (Q4.30)
39. (a) (Q4.48) Power in the sidebands = 0.5 x m^2 x carrier power = 0.5 x 0.81 x 750 = 303.75 W. The modulator and the final RF power amplifier have efficiencies of 25 percent and 85 percent, respectively. The dc power input to the modulator = (303.75)(100/25)(100/85) = 1429 W.
40. (c) (R&R73.258)
41. (e) (Q4.20) db = 10 $\log_{10}$ (P gain) = 10 $\log_{10}$ (50) = 10 x 1.699 = 16.99 db.
42. (e) (Q4.89 and R&R73.682)
43. (b) (Q4.92 and R&R73.39)
44. (d) (Q4.78 and Q4.79)
45. (c) Directional antennas are required by a station to protect the service area of another higher class station.
46. (d) (Q4.123 and R&R73.681)
47. (d) (Q4.96 and R&R73.52)
48. (a) (Q4.76 and R&R73.681)
49. (c) (Q4.89 and R&R73.682)
50. (e) (Q4.25 and Q4.26)

Element 4, Test 9 Answers

1. (e) (Q4.105 and R&R73.98)
2. (a) In the interval of time for the wave of the north antenna to arrive at the south antenna, the current in the south antenna will have changed by 90° and will be out of phase and the two waves will cancel each other. The wave produced by the south antenna will arrive at the north antenna in phase and will add to the radiation in the north direction, producing the cardioid pattern with the main lobe toward the north.
3. (c) (Q3.163)
4. (c) (Q4.64 and Q3.203) Antenna field gain = $\sqrt{50 \text{ kW}/12 \text{ kW}} = 2.04$.
5. (b) (R&R2.201)
6. (d) (Q3.147 and Q4.48) Assume a carrier power of 1000 Watts. The power in the sidebands is found from the formula, $P_{sb} = (m^2/2) \times P_c$, where m = degree of modulation and P_c = carrier power. Thus, $P_{sb} = (1^2/2) \times 1000 = 500$ Watts. Total radiated power = $P_c + P_{sb} = 1000 + 500 = 1500$ W. Percentage of total power (P_T) in sidebands = $(P_{sb}/P_T) \times 100 = (500/1500) \times 100 = 33\ 1/3\%$.
7. (a) (Q4.74)
8. (b) (Q4.53) A recycling relay is the type that, when tripped by a temporary overload, will reset when the overload ceases.
9. (c) (Q3.72) Output voltage at full load (V_{FL}) = $V_{NL}/[1 + (R/100)]$ where R = regulation percentage = $230/(1 + 0.22)$ = 189 V, approximately.
10. (c) (Q4.37 and Q4.38)
11. (b) (Q3.203 and Q4.69)
12. (b) (Q3.34) For coils whose Q is greater than 10, the power factor is approximately the reciprocal of Q. Therefore Q = 1/0.02 = 50.
13. (e) (Q3.163 and R&R73.254)
14. (b) The various modes of vibration of a crystal are shown in Fig. 1.
15. (e) (Q3.213 and Q3.215)
16. (d) The high frequency gain decreases to approach zero.
17. (a) (Q3.144 and Q4.46)
18. (e) (Q3.218 and Q4.87)
19. (d) (Q4.55)
20. (c) (R&R73.52)
21. (c) A change in the value of C4 would not change the output frequency of the transmitter, but the output amplitude of Q2 may change.
22. (a) Most broadcast station monitors require a source of unmodulated RF input so that the sideband frequencies will not affect the reading of the monitor.
23. (a) (Q4.65)
24. (a) (Q4.73) The remainder of the 262.5 lines (20) are blanked out during the vertical retrace interval.
25. (c) (Q4.09) The gain (A) of a triode amplifier may be found from the equation $A = \mu R_L/(r_p + R_L) = (12 \times 30 \times 10^3)/[(7.5 \times 10^3) + (30 \times 10^3)] = 9.6$
26. (d) (Q4.30)
27. (c) (Q4.86) Because of the very wide bandwidth involved, the circuit impedances are necessarily low. With such low impedances, the amount of power required to plate modulate a television video transmitter would be prohibitive. Much less power is required to grid modulate the video transmitter.
28. (d) The load resistance presented to the modulator of a triode class "C" RF amplifier may be found by Ohm's law, R = E/I.
29. (a) (Q3.53 and Q4.08)
30. (b) (Q4.115 and R&R73.293)
31. (a) The voltage drop across the 4.7 kohms resistor is 15 V - 6.5 V = 8.5 V and the collector current is 8.5 V/4.7 kohms = 1.81 mA. With the static current gain equal to unity, I_E = 1.81 mA also, and the value of R_E is $(12.0 - 0.7)/1.81$ = 6.2 kohms, approximately.
32. (a) (Q3.34) Since $Q = 2\pi f_r L/R$, halving L will halve Q, provided f_r and R do not change.
33. (c) (Q4.86)
34. (d) (Q4.121 and R&R73.668)

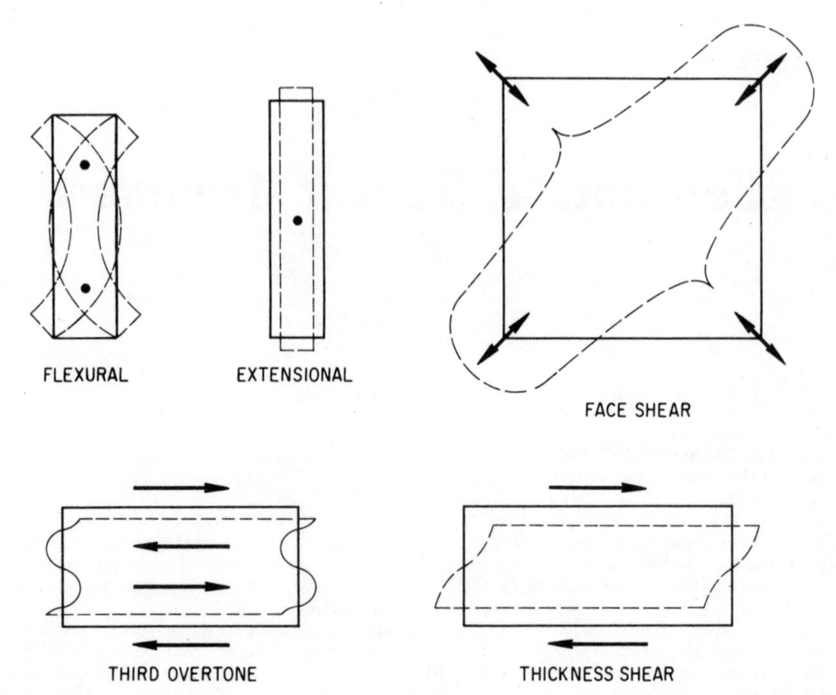

FIGURE 1

35. (a) (Q4.51)
36. (b) (Q3.210)
37. (c) (Q3.177 and Q4.59)
38. (c) (Q4.34, Q4.135, and R&R74.501)
39. (a) (Q3.187 and 3.189)
40. (b) (Q4.34, Q4.135, and R&R74.501)
41. (b) (Q4.89 and R&R73.682)
42. (e) (Q3.86)
43. (e) (Q3.155) The frequencies that will appear in the output of the mixer are the two fundamental frequencies and their sum and difference frequencies. 10.7 MHz has nothing in common with the frequencies that would appear in the output of the mixer.
44. (c) (R&R73.569)
45. (c) (R&R73.40)
46. (e) (Q4.78 and R&R73.685)
47. (a) The voltage gain is +1 in this voltage follower circuit. The EI input signal is not inverted.
48. (c) (Q4.81)
49. (b) (Q4.81)
50. (a) (Q4.81)

Element 4, Test 10 Answers

1. (e) (Q4.40)
2. (d) There would be no drive to Q2, and it would not conduct. The potential between the emitter and the collector would be 12 V dc.
3. (e) (Q4.133, R&R74.435, and R&R74.534)
4. (c) To erase the tape thoroughly in order to obtain a very low background noise, a sine wave is used at a frequency of 60 to 100 kHz.
5. (c) (Q4.58)
6. (a) (Q4.76 and R&R73.699)
7. (e) (Q4.76 and R&R73.699)
8. (c) (Q4.76 and R&R73.699)
9. (c) (Q4.76 and R&R73.699)
10. (c) (Q4.76 and R&R73.699)
11. (e) (Q4.93 and R&R73.40)
12. (e) Tape dropout is defined as a loss of signal from the tape due to lack of oxide on the tape.
13. (e) (Q3.84 and Q4.83)
14. (b) The "Austin ring" is a transformer that provides a low impedance path for the 60 Hz power feeding the tower lights and a high impedance path to block the RF from the power line.
15. (b) (Q4.80 and Q4.81)
16. (c) (Q4.64) RF power input to the antenna = $(3\text{ A})^2 \times 22$ ohms = 198 W. DC power input to the final power amplifier = 600 V x 1/2 A = 300 W. RF power output from the final power amplifier = 300 W x (75/100) = 225 W. Power loss between the output stage and the antenna = 225 - 198 = 27 W.
17. (e) (Q4.77)
18. (b) (Q4.111 and R&R73.262)
19. (b) (Q4.117 and R&R73.310)
20. (b) (Q4.104 and R&R73.60)
21. (e) (Q4.145 and R&R17.50)
22. (e) (Q4.124 and R&R73.689)
23. (d) (Q4.25) The voltage gain with feedback = $A/(1 + A\beta)$, where β = 20/100 = 0.2 and A = 65. Thus, the voltage gain = $65/[1 + (65 \times 0.2)] = 65/14 = 4.6$, approximately.
24. (c) (Q3.142 and Q4.58) The percentage modulation (where $E_{av} = (14 + 2)/2 = 8$) is equal to $[(E_{max} - E_{min})/2\,E_{av}] \times 100$ = $[(14 - 2)/(2 \times 8)] \times 100 = 75\%$.
25. (b) (Q3.203) The wavelength corresponding to 1245 kHz (1.245 MHz) = (2 x 468)/1.245 = 751.8 ft. The physical height of the tower = 751.8 x 0.65 = 488.7 ft.
26. (c) Of the antennas listed, the most popular is the stacked ringed dipole. Its polarization is both horizontal and vertical, providing better reception for mobile car receivers and whip antennas used with portable radios.
27. (d) The frequency is given by:
    $f = (0.159)/(C \times X_c) = (0.159)/(2 \times 10^{-12} \times 10{,}600)$
    $= (0.159)/(2 \times 10^{-12} \times 10.6 \times 10^3)$
    $= (0.159)/(21.2 \times 10^{-9})$
    $= (159 \times 10^6)/(21.2) = 7.5$ MHz
28. (e) (Q4.71 and Q4.72)
29. (d) (Q4.117, Q3.203(a), and Q4.57) Field strength is directly proportional to the square root of the transmitter power. Therefore, new transmitter power/15 kW = $(75/50)^2$. Solving, new transmitter power = 15 kW x 2.25 = 33.75 kW.
30. (b) (Q4.90) The current flowing in the transmission line (I) can be found by the formula $I = \sqrt{P/R}$, where I is in amps, P is power, in watts, and R is the surge impedance of the transmission line. The current in the line is $\sqrt{P/R} = \sqrt{(10 \times 10^3)/72}$ $= \sqrt{1.389 \times 10^2} = 11.8$ A, approximately.
31. (a) (Q3.192A) Collector current = Emitter current - Base current = 5 mA - 100 μA = 4.9 mA. Current gain = I_c/I_e = 4.9 mA/5 mA = 0.98.
32. (d) (Q3.143, Q3.291, and Q4.46)
33. (c) The value of the screen grid resistance may be found by Ohm's law, using the voltage drop across the resistance and the current through it. The voltage drop across R_{sg} = 2000 V - 350 V = 1.65 kV. $R_{sg} = V/I = (1.65 \times 10^3)/0.040 = 41.25$ kohms.
34. (b) The required secondary impedance of the modulation transformer may be found by Ohm's law. Z = V/I, where Z = the secondary impedance of the transformer,

in ohms, V = the plate supply voltage, in volts, and I = the plate current plus the screen current, in amps. Z = 2000/0.24 = 8.333 kohms.

35. (b) (Q3.59 and Q4.102) The dc input power to the plate and screen circuit = 2000 V x 240 mA = 480 W. The plate efficiency = (RF output power/dc input power) x 100 = (300 W/480 W) x 100 = 62.5 percent.
36. (a) The emitter current = 9 V/12 kohms = 0.75 mA. The voltage drop across the collector resistor = 10 k x 0.75 mA = 7.5 V. The collector potential = 12 V - 7.5 V = +4.5 V.
37. (b) (Q4.140 and R&R17.25)
38. (c) (Q3.86)
39. (c) (Q4.66)
40. (e) (Q4.101 and R&R73.47)
41. (c) (Q4.112 and R&R73.265)
42. (e) (Q4.40 and Q4.41)
43. (b) (Q4.90 and R&R73.3)
44. (a) (Q4.109 and R&R73.116)
45. (b) (Q4.20 and Q4.64) The ratio of the voltage received at the fundamental to the voltage received at the harmonic can be used in the formula, $Ndb = 20 \log_{10} (E_2/E_1)$. $Ndb = 20 \log_{10} (150 \times 10^{-3})/(150 \times 10^{-5}) = 20 \log_{10} (1 \times 10^2) = 20 \times 2 =$ 40 dB.
46. (d) (R&R73.55 and Q4.103)
47. (b) (Q4.89)
48. (a) (Q4.35 and R&R73.254)
49. (b) (Q4.81)
50. (a) (Q4.84)

Answer Sheets

Element 4 Test 1 Answer Sheet	*Element 4 Test 2 Answer Sheet*	*Element 4 Test 3 Answer Sheet*	*Element 4 Test 4 Answer Sheet*	*Element 4 Test 5 Answer Sheet*
1. (a)	1. (d)	1. (d)	1. (b)	1. (d)
2. (e)	2. (c)	2. (c)	2. (e)	2. (b)
3. (e)	3. (c)	3. (c)	3. (c)	3. (d)
4. (c)	4. (c)	4. (a)	4. (c)	4. (d)
5. (c)	5. (c)	5. (c)	5. (c)	5. (b)
6. (c)	6. (d)	6. (c)	6. (b)	6. (b)
7. (d)	7. (c)	7. (b)	7. (e)	7. (a)
8. (b)	8. (a)	8. (a)	8. (e)	8. (d)
9. (c)	9. (a)	9. (e)	9. (d)	9. (e)
10. (e)	10. (b)	10. (a)	10. (e)	10. (c)
11. (e)	11. (c)	11. (b)	11. (a)	11. (b)
12. (e)	12. (d)	12. (a)	12. (b)	12. (b)
13. (a)	13. (a)	13. (d)	13. (e)	13. (e)
14. (b)	14. (b)	14. (e)	14. (e)	14. (b)
15. (d)	15. (e)	15. (e)	15. (a)	15. (b)
16. (b)	16. (e)	16. (c)	16. (e)	16. (b)
17. (c)	17. (e)	17. (b)	17. (c)	17. (e)
18. (d)	18. (a)	18. (e)	18. (e)	18. (b)
19. (b)	19. (b)	19. (a)	19. (c)	19. (b)
20. (d)	20. (c)	20. (a)	20. (b)	20. (a)
21. (d)	21. (b)	21. (d)	21. (e)	21. (a)
22. (e)	22. (c)	22. (c)	22. (d)	22. (e)
23. (b)	23. (e)	23. (e)	23. (a)	23. (b)
24. (b)	24. (e)	24. (e)	24. (c)	24. (d)
25. (e)	25. (b)	25. (e)	25. (c)	25. (d)
26. (d)	26. (b)	26. (e)	26. (e)	26. (d)
27. (a)	27. (c)	27. (a)	27. (b)	27. (d)
28. (d)	28. (c)	28. (c)	28. (c)	28. (d)
29. (a)	29. (e)	29. (b)	29. (e)	29. (d)
30. (e)	30. (a)	30. (a)	30. (e)	30. (c)
31. (e)	31. (b)	31. (c)	31. (a)	31. (e)
32. (e)	32. (e)	32. (a)	32. (b)	32. (a)
33. (d)	33. (a)	33. (c)	33. (e)	33. (e)
34. (c)	34. (d)	34. (d)	34. (d)	34. (b)
35. (c)	35. (e)	35. (e)	35. (e)	35. (a)
36. (d)	36. (a)	36. (b)	36. (d)	36. (b)
37. (c)	37. (c)	37. (c)	37. (a)	37. (c)
38. (e)	38. (b)	38. (c)	38. (c)	38. (d)
39. (d)	39. (d)	39. (d)	39. (a)	39. (e)
40. (b)	40. (b)	40. (c)	40. (b)	40. (b)
41. (d)	41. (d)	41. (c)	41. (b)	41. (d)
42. (a)	42. (b)	42. (c)	42. (d)	42. (a)
43. (a)	43. (c)	43. (c)	43. (e)	43. (c)
44. (c)	44. (b)	44. (a)	44. (c)	44. (b)
45. (b)	45. (e)	45. (e)	45. (e)	45. (c)
46. (a)	46. (a)	46. (e)	46. (c)	46. (b)
47. (b)	47. (d)	47. (c)	47. (c)	47. (d)
48. (a)	48. (c)	48. (b)	48. (e)	48. (e)
49. (c)	49. (c)	49. (d)	49. (d)	49. (c)
50. (c)	50. (d)	50. (c)	50. (c)	50. (d)

	Element 4 Test 6 Answer Sheet	*Element 4 Test 7 Answer Sheet*	*Element 4 Test 8 Answer Sheet*	*Element 4 Test 9 Answer Sheet*	*Element 4 Test 10 Answer Sheet*
1.	(e)	(d)	(d)	(e)	(e)
2.	(c)	(a)	(d)	(a)	(d)
3.	(e)	(b)	(c)	(c)	(e)
4.	(e)	(b)	(a)	(c)	(c)
5.	(d)	(d)	(e)	(b)	(c)
6.	(b)	(b)	(e)	(d)	(a)
7.	(b)	(e)	(c)	(a)	(e)
8.	(e)	(c)	(e)	(b)	(c)
9.	(a)	(c)	(e)	(c)	(c)
10.	(b)	(e)	(c)	(c)	(c)
11.	(a)	(c)	(d)	(b)	(e)
12.	(d)	(c)	(a)	(b)	(e)
13.	(c)	(a)	(a)	(e)	(e)
14.	(a)	(d)	(e)	(b)	(b)
15.	(b)	(b)	(a)	(e)	(b)
16.	(e)	(d)	(a)	(d)	(c)
17.	(a)	(e)	(b)	(a)	(e)
18.	(a)	(b)	(a)	(e)	(b)
19.	(c)	(b)	(b)	(d)	(b)
20.	(b)	(e)	(c)	(c)	(b)
21.	(b)	(a)	(a)	(c)	(e)
22.	(d)	(a)	(e)	(a)	(e)
23.	(d)	(d)	(c)	(a)	(d)
24.	(c)	(d)	(e)	(a)	(c)
25.	(e)	(d)	(c)	(c)	(b)
26.	(e)	(c)	(d)	(d)	(c)
27.	(b)	(b)	(c)	(c)	(d)
28.	(e)	(d)	(e)	(d)	(e)
29.	(b)	(d)	(b)	(a)	(d)
30.	(d)	(e)	(a)	(b)	(b)
31.	(a)	(e)	(c)	(a)	(a)
32.	(c)	(c)	(b)	(a)	(d)
33.	(d)	(a)	(e)	(c)	(c)
34.	(c)	(a)	(c)	(d)	(b)
35.	(c)	(a)	(e)	(a)	(b)
36.	(a)	(b)	(c)	(b)	(a)
37.	(d)	(a)	(a)	(c)	(b)
38.	(e)	(c)	(c)	(c)	(c)
39.	(b)	(b)	(a)	(a)	(c)
40.	(b)	(d)	(c)	(b)	(e)
41.	(a)	(e)	(e)	(b)	(c)
42.	(e)	(c)	(e)	(e)	(e)
43.	(d)	(a)	(b)	(e)	(b)
44.	(c)	(a)	(d)	(c)	(a)
45.	(e)	(a)	(c)	(c)	(b)
46.	(e)	(b)	(d)	(e)	(d)
47.	(a)	(c)	(d)	(a)	(b)
48.	(e)	(a)	(a)	(c)	(a)
49.	(e)	(c)	(c)	(b)	(b)
50.	(b)	(b)	(e)	(a)	(a)